19th International Conference on Recent Progress in Many-Body Theories 2017 (MBT19)

Journal of Physics: Conference Series Volume 1041

Pohang, South Korea
25 - 30 June 2017

Editors:

Jong E. Han
Ki-Seok Kim

ISBN: 978-1-5108-6492-4
ISSN: 1742-6588

Printed from e-media with permission by:

Curran Associates, Inc.
57 Morehouse Lane
Red Hook, NY 12571

CURRAN ASSOCIATES INC. proceedings.com

Printed by Curran Associates, Inc. (2018)

Institute of Physics
Dirac House, Temple Back
Bristol BS1 6BE UK

Phone: 44 1 17 929 7481
Fax: 44 1 17 920 0979

techtracking@iop.org

Additional copies of this publication are available from:

Curran Associates, Inc.
57 Morehouse Lane
Red Hook, NY 12571 USA
Phone: 845-758-0400
Fax: 845-758-2633
Email: curran@proceedings.com
Web: www.proceedings.com

Table of contents

Volume 1041

19th International Conference Recent Progress in Many-Body Theories (RPMBT)

25–30 June 2017, Pohang, South Korea

Accepted papers received: 24 May 2018
Published online: 12 June 2018

Preface

Papers

Condensed Matter Physics

19th International Conference on Recent Progress in Many-Body Theories (MBT19)

Jong E. Han[1], **Ki-Seok Kim**[2]

[1] Department of Physics, University at Buffalo SUNY, Buffalo NY 14260, USA

[2] Department of Physics, Pohang University of Science and Technology, Pohang, Korea

E-mail: jonghan@buffalo.edu

The 19th International Conference on Recent Progress in Many-Body Theories (MBT19) was held at the Asia Pacific Center for Theoretical Physics (APCTP) on June 25-30, 2017. The conference continued the approximately biennial series initiated in Trieste in 1978 and was devoted to new developments in the field of many-body theories, with an added focus on electron transport and nanostructures. The conference series encourages the exchange of ideas between physicists working in such diverse areas as nuclear and condensed matter physics, quantum chemistry, lattice Hamiltonians or quantum fluids. This volume includes the proceedings of the conference.

Many-body theories are an integral part in different fields of theoretical physics such as condensed matter, nuclear matter that share common methodology often leading to unifying concepts. Phase transitions and macroscopic quantum effects such as magnetism, Bose-Einstein condensation, superfluidity or superconductivity have been investigated within ultra-cold gases, finite systems or various nanomaterials. Traditional equilibrium theories are now expanded to strong-field and ultrafast dynamics of condensed matter systems. The conference series on Recent Progress in Many-Body Theories (RPMBT) is devoted to foster interaction and cross-fertilization between different fields and also to discuss future lines of research.

The main conference program was split into six sections:

- Many-Body Problems in Condensed Matter Physics
- Quantum Fluids and Cold Gases
- Nuclear Many-Body and Relativistic Theories
- Computational Physics
- Electron Transport and Nanostructures
- New Developments in Computational Many-Body Techniques

Sixty four participants from eighteen countries participated. Sixty four contributions were presented, and forty nine of them were invited talks. We were fortunate to have outstanding speakers which were able to convey the excitement of these fields, and would like to thank all them for making this conference so lively. As a particular highlight of the conference, the Eugene Feenberg Memorial Medal for outstanding results in the field of many-body theory, and the Hermann Kümmel Early Achievement Award in Many-Body Physics for young scientists in that field were awarded.

The Eugene Feenberg Medal was awarded to Jordi Boronat. The citation reads as follows: For his development and implementation of microscopic many-body methods, both analytical and computational, and for his use of them to accurately determine properties of quantum fluids and dilute Fermi gases in various geometries and in the BEC-BCS crossover regime.

The Hermann Kümmel Early Achievement Award in Many-Body Physics was awarded to Guy Cohen: For the development and implementation of novel algorithms for nonequilibrium and equilibrium quantum many-body physics.

The Conference continues the series of conferences held before in Trieste, Italy (1978); Oaxtapec, Mexico (1981); Odenthal-Altenberg, Germany (1983); San Francisco, USA (1985); Oulu, Finland (1987); Arad, Israel (1989); Minneapolis, USA (1991); Schlo Segau, Austria (1994); Sydney, Australia (1997); Seattle, USA (1999); Manchester, UK (2001); Santa Fe, USA (2004); Buenos Aires, Argentina (2005); Barcelona, Spain (2007); Columbus, USA (2009); Bariloche, Argentina (2011); Rostock, Germany (2013) and Buffalo, USA (2015).

We thank the International Advisory Committee, the International Programme Committee, and the Local Organizing Committee for their great support and advice. The next RPMBT meeting will take place in Toulouse France in September 2019.

The conference major funding, aside from registration fees, was provided by the APCTP and the Basic Science Institute of Korea.

Jong Han and Ki-Seok Kim
Guest editors

IOP Conf. Series: Journal of Physics: Conf. Series **1041** (2018) 011002 doi:10.1088/1742-6596/1041/1/011002

Peer review statement

All papers published in this volume of *Journal of Physics: Conference Series* have been peer reviewed through processes administered by the proceedings Editors. Reviews were conducted by expert referees to the professional and scientific standards expected of a proceedings journal published by IOP Publishing.

Low-energy parameters and spin gap of a frustrated spin-s Heisenberg antiferromagnet with $s \le \frac{3}{2}$ on the honeycomb lattice

R F Bishop and P H Y Li

School of Physics and Astronomy, Schuster Building, The University of Manchester, Manchester, M13 9PL, UK

School of Physics and Astronomy, University of Minnesota, 116 Church Street SE, Minneapolis, Minnesota 55455, USA

E-mail: raymond.bishop@manchester.ac.uk; peggyhyli@gmail.com

Abstract. The coupled cluster method is implemented at high orders of approximation to investigate the zero-temperature ($T = 0$) phase diagram of the frustrated spin-s J_1–J_2–J_3 antiferromagnet on the honeycomb lattice. The system has isotropic Heisenberg interactions of strength $J_1 > 0$, $J_2 > 0$ and $J_3 > 0$ between nearest-neighbour, next-nearest-neighbour and next-next-nearest-neighbour pairs of spins, respectively. We study it in the case $J_3 = J_2 \equiv \kappa J_1$, in the window $0 \le \kappa \le 1$ that contains the classical tricritical point (at $\kappa_{\rm cl} = \frac{1}{2}$) of maximal frustration, appropriate to the limiting value $s \to \infty$ of the spin quantum number. We present results for the magnetic order parameter M, the triplet spin gap Δ, the spin stiffness ρ_s and the zero-field transverse magnetic susceptibility χ for the two collinear quasiclassical antiferromagnetic (AFM) phases with Néel and striped order, respectively. Results for M and Δ are given for the three cases $s = \frac{1}{2}$, $s = 1$ and $s = \frac{3}{2}$, while those for ρ_s and χ are given for the two cases $s = \frac{1}{2}$ and $s = 1$. On the basis of all these results we find that the spin-$\frac{1}{2}$ and spin-1 models both have an intermediate paramagnetic phase, with no discernible magnetic long-range order, between the two AFM phases in their $T = 0$ phase diagrams, while for $s > 1$ there is a direct transition between them. Accurate values are found for all of the associated quantum critical points. While the results also provide strong evidence for the intermediate phase being gapped for the case $s = \frac{1}{2}$, they are less conclusive for the case $s = 1$. On balance however, at least the transition in the latter case at the striped phase boundary seems to be to a gapped intermediate state.

1. Introduction

Extended, uniform spin-lattice models of quantum magnets comprise a number $N(\to \infty)$ of SU(2) spins with a given spin quantum number s placed on the sites of a specified regular, periodic lattice in d dimensions. The interactions between the spins are typically modelled by pairwise Heisenberg exchange interactions with specified coupling strengths. Of particular interest in this context are models that exhibit both quasiclassical behaviour and nonclassical behaviour, typically with paramagnetic phases in some part of the relevant parameter space that do not exist for their classical counterparts (i.e., with $s \to \infty$), and which hence do not show magnetic long-range order (LRO).

Clearly, it is therefore especially advantageous to investigate models and situations in which quantum effects are enhanced. Broadly speaking, one expects that quantum fluctuations will be greater in such spin-lattice systems where each of the spin quantum number s, the dimensionality d, and the coordination number z of the lattice take lower values. However, the well-known Mermin-Wagner theorem [1] prohibits all forms of magnetic LRO for systems with $d = 1$, even at zero temperature ($T = 0$), and for systems with $d = 2$ and $T \neq 0$, since for all such systems it is not possible to break a continuous symmetry, and all such quasiclassical states with magnetic LRO break both time-reversal symmetry and SU(2) spin-rotation symmetry. Hence, two-dimensional (2D) models at $T = 0$ have come to occupy a key role in the study of quantum phase transitions (QPTs) [2, 3].

Since the honeycomb lattice has the lowest coordination number ($z = 3$) of all eleven 2D Archimedean lattices (i.e., those that comprise only regular polygons, possibly of different sorts, and with all sites equivalent to one another), spin-lattice models based on it have been the subject of intense interest in recent years. Prototypical such models that have been investigated are those in which the Hamiltonian includes only terms in which the spins at lattice sites i and j interact via an isotropic Heisenberg interaction of the form $J_{ij}\mathbf{s}_i \cdot \mathbf{s}_j$. Much attention has been focussed on the case when the exchange couplings J_{ij} are restricted to be between nearest-neighbour (NN) pairs, all with equal strength J_1, next-nearest-neighbour (NNN) pairs, all with equal strength J_2, and next-next-nearest-neighbour (NNNN) pairs, all with equal strength J_3. Both the resulting so-called J_1–J_2–J_3 model and, particularly, the two special cases of it with $J_3 = 0$ and $J_3 = J_2$ have been extensively investigated with a wide variety of theoretical techniques. While frustrated such honeycomb lattice models with spins having $s = \frac{1}{2}$ have been particularly intensively studied [4–29], considerably less attention has been paid to their counterparts with $s > \frac{1}{2}$ [30–34].

Our main aim in the present work is to extend the study of such honeycomb-lattice models with $s \geq 1$, by applying to them the coupled cluster method (CCM) (see, e.g., Refs. [35–38]) implemented to high orders of approximation. The CCM has been shown to give results of unsurpassed accuracy to an extremely wide array of physical systems, including those in condensed matter physics, quantum chemistry, atomic and molecular physics, quantum optics and solid-state optoelectronics, and nuclear and subnuclear physics (and see, e.g., Refs. [35–56] and references cited therein). We note in particular that the CCM has already been applied with great success to a wide diversity of spin-lattice problems of interest in quantum magnetism (and see, e.g., Refs. [13, 19–23, 29, 32–34, 53–56] and references cited therein). These include specific applications [13, 21, 29, 33] to the model that we study further here, as discussed in more detail in Sec. 2.

The plan for the reminder of this paper is as follows. In Sec. 2 we first describe the model, including some of the pertinent results for both the classical ($s \to \infty$) and $s = \frac{1}{2}$ cases. The CCM is then briefly reviewed in Sec. 3, before we present in Sec. 4 our results using it for the particular cases $s = 1$ and $s = \frac{3}{2}$. In particular, we present results for the ground-state (GS) magnetic order parameter, the triplet spin gap, the GS spin stiffness and the GS zero-field transverse magnetic susceptibility. Finally, we conclude with a discussion of the results in Sec. 5.

2. The model

The Hamiltonian of the J_1–J_2–J_3 model on the honeycomb lattice is specified as

$$\begin{aligned} H &= J_1 \sum_{\langle i,j \rangle} \mathbf{s}_i \cdot \mathbf{s}_j + J_2 \sum_{\langle\langle i,k \rangle\rangle} \mathbf{s}_i \cdot \mathbf{s}_k + J_3 \sum_{\langle\langle\langle i,l \rangle\rangle\rangle} \mathbf{s}_i \cdot \mathbf{s}_l \\ &\equiv J_1 h(x,y)\,; \quad x \equiv J_2/J_1\,, \quad y \equiv J_3/J_1\,, \end{aligned} \tag{1}$$

where the operators $\mathbf{s}_i \equiv (s_i^x, s_i^y, s_i^z)$ are the usual SU(2) quantum spin operators on lattice site i, with $\mathbf{s}_i^2 = s(s+1)\mathbb{1}$. We shall compare and contrast here the cases with $s = \frac{1}{2}$, $s = 1$ and

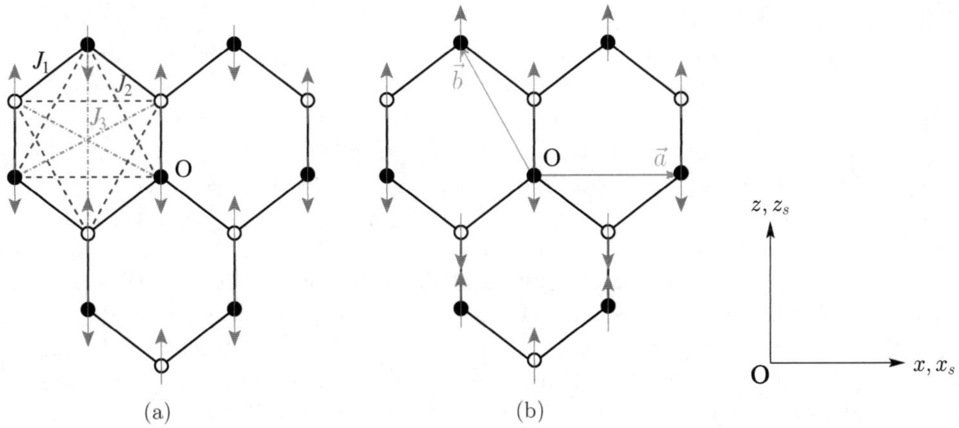

Figure 1. The J_1–J_2–J_3 honeycomb model with $J_1 > 0$; $J_2 > 0$; $J_3 > 0$, showing (a) the bonds (J_1 = —— ; J_2 = – – – ; J_3 = – · –) and the Néel state, and (b) triangular Bravais lattice vectors **a** and **b** and one of the three equivalent striped states. Sites on sublattices $\mathcal{A}$ and $\mathcal{B}$ are shown by filled and empty circles respectively, and spins on the lattice are represented by the (red) arrows on the lattice sites. We also show both the lattice-plane axes (x, z) and the spin-space axes (x_s, z_s).

$s \geq \frac{3}{2}$. The sums in Eq. (1) over $\langle i,j \rangle$, $\langle\langle i,k \rangle\rangle$ and $\langle\langle\langle i,l \rangle\rangle\rangle$ run over all NN, NNN and NNNN bonds respectively, counting each bond once only in each sum. The lattice and the exchange bonds are illustrated in Fig. 1(a). We shall be interested in the present paper in the case when all three bonds are antiferromagnetic (AFM) (i.e., when $J_i > 0$; $i = 1,2,3$). The NN exchange coupling constant J_1 simply sets the overall energy scale, and hence the nontrivial parameters of the model may be taken as $J_2/J_1 \equiv x$ and $J_3/J_1 \equiv y$, so that we may rewrite the Hamiltonian as $H \equiv J_1 h(x,y)$, as in Eq. (1).

As we discuss in more detail below, the classical version ($s \to \infty$) of the model has a tricritical point at $x = y = \frac{1}{2}$ in its $T = 0$ phase diagram [4,6]). At this point the Hamiltonian may be rewritten in the particularly simple form

$$h(\tfrac{1}{2}, \tfrac{1}{2}) = \frac{1}{4} \sum_{\chi} \mathbf{s}_{\chi}^2 - \frac{3}{4} N s(s+1)\,, \tag{2}$$

of a sum over edge-sharing hexagonal plaquettes χ, up to an additive constant term, where

$$\mathbf{s}_{\chi} = \sum_{i \in \chi} \mathbf{s}_i \tag{3}$$

is the total spin of the six spins on the elementary hexagon χ. Clearly, at this special point $x = y = \frac{1}{2}$, any state wth zero total spin on each elementary hexagonal plaquette is a classical $T = 0$ GS phase, thereby giving rise to a macroscopic degeneracy. In fact, as we describe below, one of the three classical states with perfect magnetic LRO that emerges from this tricritical point, is itself also associated with an infinitely degenerate family (IDF) of GS phases. Indeed, this is one of the main reasons to be interested in the model, since classical spin-lattice systems that exhibit such an IDF of $T = 0$ GS phases are, *a priori*, prime candidates for the possible emergence of novel quantum phases (i.e., without classical counterparts) in the cases when the spin quantum number s is finite, since the role of quantum fluctuations is then, self-evidently, enhanced.

While the honeycomb lattice is bipartite (and hence not geometrically frustrated) it is non-Bravais. It comprises two sites per unit cell, with two triangular Bravais sublattices $\mathcal{A}$ and $\mathcal{B}$. If we define the lattice to occupy the xz plane, the basis vectors may be chosen as $\mathbf{a} = \sqrt{3}d\hat{x}$ and $\mathbf{b} = \frac{1}{2}(-\sqrt{3}\hat{x} + 3\hat{z})d$, as illustrated in Fig. 1(b), where $\hat{x}$ and $\hat{z}$ are unit vectors in the x and z directions respectively, and d is the honeycomb lattice spacing (i.e., the distance between NN sites). The unit cell i residing at position vector $\mathbf{R}_i \equiv m_i\mathbf{a} + n_i\mathbf{b}$, with $m_i, n_i \in \mathbb{Z}$, contains the two sites at $\mathbf{R}_i \in \mathcal{A}$ and $(\mathbf{R}_i + d\hat{z}) \in \mathcal{B}$. The corresponding reciprocal lattice vectors are hence given by $\boldsymbol{\alpha} = 2\pi(\sqrt{3}\hat{x} + \hat{z})/(3d)$ and $\boldsymbol{\beta} = 4\pi/(3d)\hat{z}$. Thus, the first Brillouin zone and the Wigner-Seitz unit cell are the respective parallelograms formed by the pairs of vectors $(\boldsymbol{\alpha}, \boldsymbol{\beta})$ and $(\mathbf{a}, \mathbf{b})$. Clearly, one may also take both as being centred on a point of sixfold rotational symmetry in their corresponding lattice planes. Thus, just as the Wigner-Seitz unit cell may be chosen to be bounded by the sides of an elementary hexagon (i.e., of side length d), as in Fig. 1, so the first Brillouin zone may also be chosen as a hexagon of side length $4\pi/(3\sqrt{3}d)$, which is rotated by 90° with respect to the Wigner-Seitz hexagon. Thus, the corners of the hexagon forming the first Brillouin zone are given by the vectors

$$\mathbf{K}^{(1)} = \frac{4\pi}{3\sqrt{3}d}\hat{x}\,, \quad \mathbf{K}^{(2)} = \frac{2\pi}{3\sqrt{3}d}(\hat{x} + \sqrt{3}\hat{z})\,, \quad \mathbf{K}^{(3)} = \frac{2\pi}{3\sqrt{3}d}(-\hat{x} + \sqrt{3}\hat{z})\,, \tag{4}$$

with the remaining three corners at positions $\mathbf{K}^{(i+3)} = -\mathbf{K}^{(i)}$; $i = 1, 2, 3$. Similarly, the mid-points of the edges of the first Brillouin zone have the vectors

$$\mathbf{M}^{(1)} = \frac{\pi}{3d}(\sqrt{3}\hat{x} + \hat{z})\,, \quad \mathbf{M}^{(2)} = \frac{2\pi}{3d}\hat{z}\,, \quad \mathbf{M}^{(3)} = \frac{\pi}{3d}(-\sqrt{3}\hat{x} + \hat{z})\,, \tag{5}$$

with the remaining three midpoints at positions $\mathbf{M}^{(i+3)} = -\mathbf{M}^{(i)}$; $i = 1, 2, 3$.

Returning now to the classical ($s \to \infty$) version of the model of Eq. (1), it has been shown [4,6] that its generic stable GS phase is described by a coplanar spiral configuration of spins. This may be defined, as usual, by a wave vector $\mathbf{Q}$, together with (for the non-Bravais two-site unit cell of the model) an angle ϕ that defines the relative orientation of the two sites in the same unit cell at position vector $\mathbf{R}_i$. The two classical spins in unit cell i thus have orientations

$$\mathbf{s}_{i,\sigma} = -s[\cos(\mathbf{Q}\cdot\mathbf{R}_i + \phi_\sigma)\hat{z}_s + \sin(\mathbf{Q}\cdot\mathbf{R}_i + \phi_\sigma)\hat{x}_s]\,; \quad \sigma = \mathcal{A}, \mathcal{B}\,, \tag{6}$$

where the spin-space plane is now defined by the two orthogonal unit vectors $\hat{x}_s$ and $\hat{z}_s$, shown in Fig. 1. The two angles ϕ_σ are chosen so that $\phi_\mathcal{A} = 0$ and $\phi_\mathcal{B} = \phi$.

In the case considered here when all three bonds are AFM (i.e., with $J_1 > 0$, $x > 0$, $y > 0$), it has been shown [4, 6] that the $T = 0$ classical GS phase diagram comprises three phases that meet at the tricritical point $(x, y) = (\frac{1}{2}, \frac{1}{2})$, discussed above. If we define our lattice origin to be at the centre of the hexagonal Wigner-Seitz unit cell, one may show that one value of the spiral wave vector $\mathbf{Q}$ that minimizes the classical GS energy in this case is

$$\mathbf{Q} = \frac{2}{\sqrt{3}d}\cos^{-1}\left[\frac{(1-2x)}{4(x-y)}\right]\hat{x}\,, \tag{7}$$

together with the value $\phi = \pi$. Clearly, for Eq. (7) to be physically valid, we require $|(1-2x)/(x-y)| \leq 4$, or equivalently,

$$y \leq \frac{3}{2}x - \frac{1}{4}\,; \quad y \leq \frac{1}{2}x + \frac{1}{4}\,. \tag{8}$$

The two straight lines defined by Eq. (8) taken as equalities cross at the (tricritical) point $(x, y) = (\frac{1}{2}, \frac{1}{2})$. Along the first boundary line, $y = \frac{3}{2}x - \frac{1}{4}$, $\mathbf{Q} = \boldsymbol{\Gamma} = (0, 0)$, which (again,

IOP Conf. Series: Journal of Physics: Conf. Series **1041** (2018) 012001 doi:10.1088/1742-6596/1041/1/012001

together with the minimizing condition $\phi = \pi$) simply describes the Néel phase illustrated in Fig. 1(a). Similarly, everywhere along the second boundary line in Eq. (8), $y = \frac{1}{2}x + \frac{1}{4}$, we have $\mathbf{Q} = 2\pi/(\sqrt{3}d)\hat{x}$. In fact this vector lies outside the first Brillouin zone. When mapped back inside it takes the equivalent value $\mathbf{Q} = \mathbf{M}^{(2)}$, one of the midpoints of the edges of the hexagonal first Brillouin zone, given in Eq. (5). This wave vector (again together with the minimizing condition $\phi = 0$) describes the striped AFM phase illustrated in Fig. 1(b). It is clear that the phase transitions across the boundaries of Eq. (8) considered as equalities (i.e., between the Néel and spiral phases, and between the striped and spiral phases) are both continuous ones. One may also readily show from energetics that the phase transition between the two collinear AFM (Néel and striped) phases is of first-order type, and occurs along the boundary $x = \frac{1}{2}$, which again meets the other two phase boundaries at the classical tricritical point $(x, y) = (\frac{1}{2}, \frac{1}{2})$.

It is obvious that both the striped phase and the spiral phase described by Eq. (7) break the rotational symmetry and that there must be two other equivalent states in each case, obtained by a rotation of $\pm\frac{2}{3}\pi$ in the honeycomb xz plane. For the striped state this threefold degeneracy is simply equivalent to the wave vector $\mathbf{Q}$ being allowed to take any of the position vectors $\mathbf{M}^{(i)}$; $i = 1, 2, 3$, of the three inequivalent midpoints of the sides of the hexagonal first Brillouin zone, as given by Eq. (5), together with $\phi = 0$ in the cases with $i = 1, 3$.

Although it is known [57] that classical GS spin configurations can generally be described as in Eq. (6), it is also known [57] that there are exceptional cases when the GS order is either not unique (up to a global rotation) or has a discrete degeneracy such as that associated with the striped state. Such exceptions are known [6, 57] to occur for special values of the ordering wave vector $\mathbf{Q}$. These include the cases when $\mathbf{Q}$ takes a value equal to either one half or one quarter of a reciprocal lattice vector $\mathbf{G}_i \equiv k_i\boldsymbol{\alpha} + l_i\boldsymbol{\beta}$, with $k_i, l_i \in \mathbb{Z}$. The striped states are precisely of this form since their wave vectors $\mathbf{Q} = \mathbf{M}^{(i)}$, $i = 1, 2, 3$, are precisely one half of corresponding reciprocal lattice vectors, i.e., $\mathbf{M}^{(1)} = \frac{1}{2}\boldsymbol{\alpha}$, $\mathbf{M}^{(2)} = \frac{1}{2}\boldsymbol{\beta}$, $\mathbf{M}^{(3)} = \frac{1}{2}(\boldsymbol{\beta} - \boldsymbol{\alpha})$. In this case it has been explicitly shown [6] that the GS ordering has an IDF of non-planar spin configurations, all degenerate in energy with the collinear striped states. It has also been shown [4, 6] that, at least in the large-s limit when lowest-order spin-wave theory (LSWT) becomes exact, quantum fluctuations lift this degeneracy in favour of the AFM striped states, which now have the lowest energy.

To conclude our discussion of the classical limit ($s \to \infty$) of the J_1–J_2–J_3 model on the honeycomb lattice in the region where $J_1 > 0$, $x \geq 0$, $y \geq 0$, we have seen that at $T = 0$ its classical GS phase diagram has three phases, each with perfect magnetic LRO. These comprise (a) a Néel AFM phase in the region $y > 0$, $0 < x < \frac{1}{6}$ and $y > \frac{3}{2}x - \frac{1}{4}$, $\frac{1}{6} < x < \frac{1}{2}$; (b) a collinear striped AFM phase in the region $y > \frac{1}{2}x + \frac{1}{4}$, $x > \frac{1}{2}$; and (c) a coplanar spiral phase in the region $0 < y < \frac{3}{2}x - \frac{1}{4}$, $\frac{1}{6} < x < \frac{1}{2}$ and $0 < y < \frac{1}{2}x + \frac{1}{4}$, $x > \frac{1}{2}$. For a further discussion of the classical $T = 0$ GS phase diagram of the model, and for other exceptional cases (including those when $\mathbf{Q}$ takes a value $\frac{1}{2}\mathbf{G}_i$ or $\frac{1}{4}\mathbf{G}_i$) the reader is referred to Refs. [6, 7, 33]. Although not of direct relevance here, the case $y = 0$ that includes a spiral phase for $x > \frac{1}{6}$, also includes a one-parameter IDF of incommensurate GS phases wherein the wave vector $\mathbf{Q}$ can orient in an arbitrary direction, with degenerate solutions along a specific contour for a given value of x (and see Refs. [6, 7] for details). Again, at the level of LSWT, this degeneracy is lifted by quantum fluctuations to give spiral order by disorder.

The most interesting region of the classical phase diagram, for reasons discussed above, includes the tricritical point and the striped phase (that is part of an IDF of GS phases). Both are sampled along the line $y = x = \kappa$ (i.e., in the region $J_1 > 0$, $J_3 = J_2 \equiv \kappa J_1$, $\kappa > 0$). Hence, for the remainder of this study we work in this regime. Since the classical tricritical point is at $\kappa_{\rm cl} = \frac{1}{2}$, we shall investigate specifically the window $0 \leq \kappa \leq 1$ of the frustration parameter.

For the case $s = \frac{1}{2}$, the model has been studied previously in the same parameter range

IOP Conf. Series: Journal of Physics: Conf. Series **1041** (2018) 012001 doi:10.1088/1742-6596/1041/1/012001

$0 \le \kappa \le 1$ [9, 13, 21, 29]. Each of these studies concurs on the finding that the classical transition at $\kappa_{\rm cl} = \frac{1}{2}$ for the limiting case $s \to \infty$ is split into two transitions for the case $s = \frac{1}{2}$, one at $\kappa_{c_1} < \frac{1}{2}$ and the other at $\kappa_{c_2} > \frac{1}{2}$. Whereas LSWT provides the relatively crude estimates $\kappa_{c_1} \approx 0.29$ and $\kappa_{c_2} \approx 0.55$, the more powerful, and potentially more accurate method of Schwinger-boson mean-field theory (SBMFT) gives the estimates $\kappa_{c_1} \approx 0.41$ and $\kappa_{c_2} \approx 0.6$ [9]. These SBMFT calculations also predict a quantum disordered phase in the intermediate region $\kappa_{c_1} < \kappa < \kappa_{c_2}$, in which a gap in the bosonic dispersion opens up. These results are broadly confirmed by high-order CCM calculations [13, 21, 29], which yield the most accurate results to date for this case. From extensive calculations of a wide variety of low-energy parameters and the triplet spin gap for the model, the best CCM estimates are $\kappa_{c_1} = 0.45 \pm 0.02$ and $\kappa_{c_2} = 0.60 \pm 0.02$ [29]. Furthermore, CCM calculations of the plaquette valence-bond crystalline (PVBC) susceptibility [13] provide powerful evidence for the intermediate paramagnetic phase to be a gapped state with PVBC order over (almost all or) the entire region. Furthermore, very recently, the CCM has also been used to study a corresponding *AA*-stacked honeycomb bilayer model [58] for the case $s = \frac{1}{2}$, where each monolayer has the same bonds as here, but now with the addition of an AFM NN interlayer coupling, $J_1^{\perp} > 0$.

In view of these interesting results for the $s = \frac{1}{2}$ J_1–J_2–J_3 model with $J_3 = J_2$ and the known results for the classical limit ($s \to \infty$), it is clearly now intriguing also to consider the cases $s = 1$ and $s \geq 1$. The only results known to us are preliminary calculations of our own [33], again using the CCM. In that earlier work we calculated only the magnetic order parameter (out of the complete set of low-energy parameters) for the two AFM quasiclassical GS phases (i.e., the Néel and striped phases), for the cases $s = 1, \frac{3}{2}, 2, \frac{5}{2}$, as well as their PVBC susceptibilities. Based only on the vanishing of the magnetic order we found that the spin-1 case also had an intermediate non-classical phase, with Néel ordering now disappearing at $\kappa_{c_1} = 0.485 \pm 0.005$ and striped order disappearing at $\kappa_{c_2} = 0.528 \pm 0.005$. Just as for the $s = \frac{1}{2}$ case [13] it was also found for the $s = 1$ case [33] that the transition at κ_{c_1} appears to be of continuous type (and hence to be a candidate for deconfined quantum criticality [59, 60], since the Landau-Ginzburg-Wilson scenario cannot hold, as discussed in more detail in Ref. [13]), while that at κ_{c_2} appears to be of first-order type. Unlike in the $s = \frac{1}{2}$ case, however, where PVBC ordering seems to occur over the entire range $\kappa_{c_1} < \kappa < \kappa_{c_2}$, for the $s = 1$ case PVBC ordering appears to be absent everywhere (or, at most, to occur over only a very small part of the region). For all cases $s > 1$ studied in Ref. [33], but based only on calculations of the magnetic order parameter of the two AFM quasiclassical phases, the quantum phase diagram appears to be similar to the classical counterpart, i.e., with a direct first-order transition from the Néel to the striped phase, but with a critical value $\kappa_c(s)$ slightly greater than $\kappa_{\rm cl} = \frac{1}{2}$. Thus, $\kappa_c(\frac{3}{2}) = 0.53 \pm 0.01$ and, for values $s > \frac{3}{2}$, $\kappa_c(s)$ appears to approach the classical value $\kappa_c(\infty) = 0.5$ monotonically as s is increased.

Our intention in the present paper is to add to these earlier preliminary findings for the model with $s > \frac{1}{2}$ by calculating for the cases $s = 1$ and $s = \frac{3}{2}$ both a complete set of GS low-energy parameters, to complement the calculations of the GS magnetic order parameter, and the triplet spin gap. In this sense we parallel the development of the $s = \frac{1}{2}$ version of the model, where the order parameter was first studied in Ref. [13], and only later were the more comprehensive calculations of a complete set of low-energy parameters and the triplet spin gap performed in Ref. [29]. Specifically, we will present results here in Sec. 4, again using high-order CCM calculations, for the spin stiffness coefficient ρ_s, the zero-field (uniform) transverse magnetic susceptibility χ, and the spin gap Δ .

3. The coupled cluster method

The CCM is nowadays regarded as providing one of the most accurate and most flexible *ab initio* techniques of modern microscopic quantum many-body theory (and see, e.g., Refs. [35–38]). The

IOP Conf. Series: Journal of Physics: Conf. Series **1041** (2018) 012001 doi:10.1088/1742-6596/1041/1/012001

method is size-extensive and size-consistent at all levels of approximate implementation. Hence, it can be utilized from the outset in the infinite system ($N \to \infty$) limit, thereby obviating the need for any finite-size scaling, such as is required by most competing methods, and hence circumventing any associated source of errors. Furthermore, the method also exactly preserves both the important Hellmann-Feynman theorem and the Goldstone linked cluster theorem at every level of approximation. These features ensure that the CCM provides accurate and self-consistent sets of results for a variety of both GS and excited-state (ES) parameters for the system under study. As is done here, the method can be implemented computationally to high orders of approximation within well-defined and well-understood truncation hierarchies, as outlined in more detail below. The results become exact as the order n of the truncation approaches infinity ($n \to \infty$), and hence the *sole* approximation ever made is to extrapolate the sequence of approximants that we calculate for any specific physical parameter.

By now the CCM has been applied to a very large number of spin-lattice systems, and we hence refer the reader to the extensive literature of such applications (and see, e.g., Refs. [29,33,34,53–56,58] and references cited therein) for full details. We content ourselves here with a very brief review of some of the most pertinent details connected with the present applications. A hallmark of the CCM is the incorporation of the quantum correlations present in the exact GS or ES wave functions via a very specific exponentiated form of correlation operator acting on an appropriately chosen model (or reference) state. For GS quantities, such as the energy and magnetic order parameter (i.e., the average local on-site magnetization) M, we use (separately) here both the Néel and striped states shown in Figs. 1(a) and 1(b), respectively, as our choices of model state. In order to calculate the lowest triplet ES energy gap Δ for the two quasiclassical phases, a spin-1 reference state is created from the above respective reference states by suitably exciting a single spin. For the calculations of the spin stiffness and magnetic susceptibility the above GS reference states have to be suitably modified by the imposition of a spin twist or a magnetic field, respectively, as we now outline in more detail.

Firstly, the spin stiffness ρ_s is a measure of the resistance of the system to an imposed rotation of the order parameter by an (infinitesimal) angle θ per unit length in a specific direction. If $E(\theta)$ is the GS energy as a function of the imposed twist, then

$$\frac{E(\theta)}{N} = \frac{E(\theta=0)}{N} + \frac{1}{2}\rho_s\theta^2 + O(\theta^4)\,, \tag{9}$$

where θ has the dimensions of inverse length. Clearly, magnetic LRO melts at the point where ρ_s vanishes. For the Néel state, whose ordering wave vector $\mathbf{Q} = \mathbf{\Gamma} = (0,0)$, clearly ρ_s is independent of the direction of the applied twist. By contrast, for the striped AFM state shown in Fig. 1(b), for which $\mathbf{Q} = 2\pi/(\sqrt{3}d)\hat{x}$, the relevant direction to apply the twist is the x direction. The corresponding twisted Néel and twisted striped states [29] are then used as CCM model states for the calculation of ρ_s in both GS phases. For the classical ($s \to \infty$) version of the AFM J_1–J_2–J_3 model under study on the honeycomb lattice, with $J_3 = J_2 \equiv \kappa J_1$, the definition of Eq. (9) easily leads to the corresponding classical values of ρ_s in the two AFM phases,

$$\rho_{s;\,\mathrm{cl}}^{\text{Néel}} = \frac{3}{4}J_1(1-2\kappa)d^2s^2\,, \tag{10}$$

and

$$\rho_{s;\,\mathrm{cl}}^{\text{striped}} = \frac{3}{4}J_1(-1+2\kappa)d^2s^2\,. \tag{11}$$

both of which are positive in the respective regions of stability of the two phases. As expected, the spin stiffness of the system vanishes precisely at the corresponding classical phase transition point, $\kappa_{\mathrm{cl}} = \frac{1}{2}$.

Secondly, to calculate the transverse magnetic susceptibility when the system is aligned in either AFM phase in the spin-space z_s direction as in Fig. 1, we now place it in the transverse magnetic field $\mathbf{h} = h\hat{x}_s$. The Hamiltonian $H = H(h = 0)$ of Eq. (1) then becomes modified to $H(h) = H(0) - h\sum_{k=1}^{N} s_k^x$, in units where the gyromagnetic ratio $g\mu_B/\hbar = 1$. The spins, which were previously aligned as in either Fig. 1(a) or Fig. 1(b), now cant at an angle $\phi = \phi(h)$ with respect to their zero-field configurations along the z_s direction. These corresponding canted Néel and canted striped states [29] are then used as CCM model states for the calculation of the respective transverse magnetic susceptibility, $\chi(h) = -N^{-1}\mathrm{d}^2E/\mathrm{d}h^2$, and its zero-field limit, $\chi = \chi(0)$, in which we are interested. In practice we calculate χ from

$$\frac{E(h)}{N} = \frac{E(h=0)}{N} - \frac{1}{2}\chi h^2 + O(h^4)\,. \tag{12}$$

for some suitably small value of h. For the classical ($s \to \infty$) version of the AFM J_1–J_2–J_3 model under study on the honeycomb lattice, with $J_3 = J_2 \equiv \kappa J_1$, it is simple to calculate the canting angle $\phi(h)$ for the two quasiclassical AFM phases. The definition of Eq. (12) thus easily yields the corresponding classical values of χ for the two AFM phases,

$$\chi_{\rm cl}^{\rm N\acute{e}el} = \frac{1}{6J_1(1+\kappa)}\,, \tag{13}$$

and

$$\chi_{\rm cl}^{\rm striped} = \frac{1}{2J_1(1+7\kappa)}\,, \tag{14}$$

where both parameters are independent of s at the classical level. We note that the two values become equal ($= \frac{1}{9}J_1^{-1}$), but nonzero, at the corresponding classical phase transition point, $\kappa_{\rm cl} = \frac{1}{2}$.

We turn now to the choice of CCM approximation scheme. Once a suitable model state has been chosen, as outlined above, the approximation simply involves the choice of which multispin-flip configurations to retain in the CCM correlation operators. A rather general such approximation scheme is the so-called SUBm–n hierarchy [54]. At a given SUBm–n level one retains all multispin-flip configurations in the CCM correlation operators that involve m or fewer spin-flips spanning a range of no more than n contiguous sites on the lattice. Each single spin-flip is defined to require the action of a spin-raising operator $s_k^+ \equiv s_k^x + s_k^y$ acting once on the model ket state (in local spin axes chosen so that on each site a passive rotation has been performed to make each spin point downwards along the negative z_s axis). Furthermore, a set of lattice sites is defined to be contiguous if every site in the set is a NN (in a specified geometry) to at least one other member of the set. Clearly, as both indices m and n become infinite, the approximation becomes exact.

For spins with spin quantum number s, the maximum number of spin-flips per site, defined as above, is $2s$. Thus, when $m = 2sn$, the SUBm–n scheme becomes equivalent to the so-called localized lattice-animal-based subsystem (LSUBn) scheme in which all multispin-flip configurations in the CCM correlation operator expansions are retained that are defined over all distinct locales (or lattice animals in the usual graph-theoretic sense) on the lattice that comprise no more than n contiguous sites. Clearly, the LSUBn $\equiv$ SUB$2sn$–n scheme is only equivalent to the SUBn–n scheme for the case $s = \frac{1}{2}$. In any such scheme we utilize the (space- and point-group) symmetries of both the system Hamiltonian and the CCM model state being used to reduce the set of independent multispin-flip configurations retained at any given order to a minimal number N_f. At a given nth level of LSUBn approximation, the number $N_f = N_f(n)$ is lowest for $s = \frac{1}{2}$ and increases sharply as a function of s. Because $N_f(n)$ also increases rapidly (typically super-exponentially) with the truncation index n, the most commonly used

CCM truncation hierarchy for spins $s > \frac{1}{2}$ is the SUBn–n scheme, and it is that scheme we employ here, just as in our previous work [33] on this model for cases with $s \geq 1$. In order to attain the higher values of the truncation index n necessary for high accuracy, we also use massively parallel supercomputing resources together with a specially tailored computer-algebra package [61] to derive and solve [53] the corresponding sets of CCM bra- and ket-state equations for both GS and ES quantities in the SUBn–n truncation scheme.

For the calculation of both the GS order parameter M and the ES gap parameter Δ we are restricted to SUBn–n calculations with $n \leq 12$ for the $s = \frac{1}{2}$ model and with $n \leq 10$ for the corresponding models with $s \geq 1$. For example, at the LSUB12 level for the calculation of M in the case $s = \frac{1}{2}$, we have $N_f(12) = 103\,097$ $(250\,891)$ when the Néel (striped) state is used as the CCM model state. For comparison, at the SUB10–10 level for calculations of M, we have corresponding numbers of fundamental configurations $N_f(10) = 219\,521$ $(552\,678)$ for the $s = 1$ model and $N_f(10) = 461\,115$ $(1\,207\,202)$ for the corresponding $s = \frac{3}{2}$ model. Similarly at the LSUB12 level for the calculation of Δ for the spin-$\frac{1}{2}$ model, the corresponding numbers are $N_f(12) = 182\,714$ $(465\,196)$ based on the Néel (striped) state as CCM model state. For comparison, at the SUB10-10 level for calculations of Δ, we have $N_f(10) = 244\,533$ $(642\,054)$ for the spin-1 model and $N_f(10) = 418\,164$ $(1\,123\,343)$ for the spin-$\frac{3}{2}$ model.

Due to the considerably reduced symmetries of both the twisted and canted model states in both quasiclassical phases, the corresponding SUBn–n calculations of ρ_s and χ can only be performed at orders $n \leq 10$ for the case $s = \frac{1}{2}$ and $n \leq 8$ for the cases $s \geq 1$. Thus, for example, at the LSUB10 level for the calculation of ρ_s in the case $s = \frac{1}{2}$, we have $N_f(10) = 347\,287$ when either the twisted Néel or the twisted striped state is used as the CCM model state. The corresponding number for ρ_s for the spin-1 model is $N_f(8) = 352\,515$ at the SUB8–8 level. Similarly, at the LSUB10 level for the calculation of χ for the spin-$\frac{1}{2}$ model, we have $N_f(10) = 58\,537$ $(174\,692)$ when the canted Néel (canted striped) state is used as the CCM model state. Corresponding numbers at the SUB8–8 level for χ for the spin-1 model are $N_f(8) = 59\,517$ $(177\,331)$.

As we have noted previously, once we have calculated CCM SUBn–n approximants for any physical parameter the *only* approximation that we ever make is the extrapolation to the (in principle) exact limit, $n \to \infty$. Although no exact such schemes are known theoretically, by now a great deal of practical experience has been built up from many applications to diverse models, so that we now have a uniform set of simple extrapolation rules, one for each parameter, that are applied consistently. For example, for highly frustrated spin-lattice models, particularly in cases where the system is close to a QCP or where the magnetic order parameter M is either zero or close to zero, a well-tested scaling scheme for M has been found (and see, e.g., Refs. [13, 19–23, 33, 34, 55, 56]) to be

$$M(n) = \mu_0 + \mu_1 n^{-1/2} + \mu_2 n^{-3/2}\,. \tag{15}$$

Similar such schemes for the spin gap, spin stiffness and zero-field transverse magnetic susceptibility all have a leading exponent of -1 (and see, e.g., Refs. [34, 56]),

$$\Delta(n) = d_0 + d_1 n^{-1} + d_2 n^{-2}\,, \tag{16}$$

$$\rho_s(n) = s_0 + s_1 n^{-1} + s_2 n^{-2}\,, \tag{17}$$

and

$$\chi(n) = x_0 + x_1 n^{-1} + x_2 n^{-2}\,. \tag{18}$$

Clearly, in order to extract the corresponding extrapolants μ_0 (for M), d_0 (for Δ), s_0 (for ρ_s) and x_0 (for χ) from Eqs. (15)–(18), respectively, we need to input at least three different corresponding SUBn–n approximants. Further details on the choice of such SUBn–n input sets are given in Sec. 4.

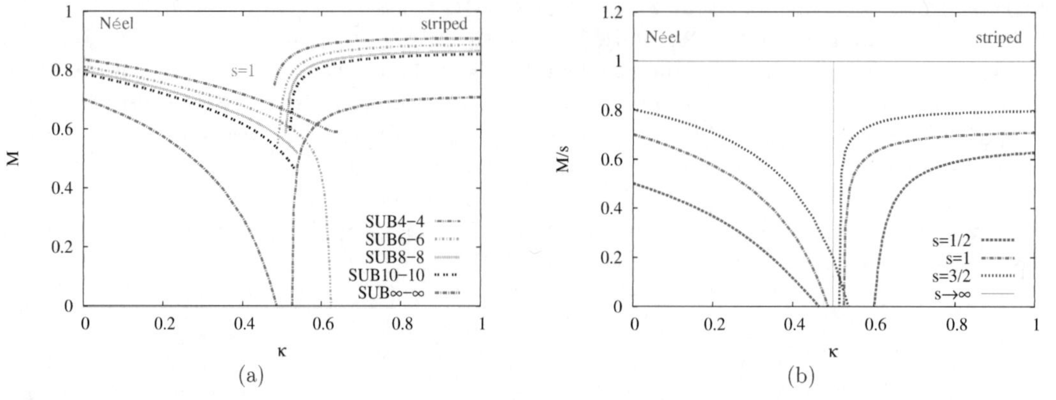

Figure 2. CCM results for the GS magnetic order parameter M for the J_1–J_2–J_3 model on the honeycomb lattice, with $J_1 > 0$ and $J_3 = J_2 \equiv \kappa J_1 > 0$, as a function of the frustration parameter κ, using both the Néel and striped states as the CCM model state. (a) Results are shown for the $s = 1$ case in SUBn–n approximations with $n = 4, 6, 8, 10$, together with the respective SUB∞–∞ extrapolant based on Eq. (15) and using this data set as input. (b) Extrapolated (SUB∞–∞) results for M/s are shown for the three cases $s = \frac{1}{2}, 1, \frac{3}{2}$. In each case Eq. (15) has been used, together with the corresponding SUBn–n data sets $n = \{6, 8, 10, 12\}$ for $s = \frac{1}{2}$ and $n = \{4, 6, 8, 10\}$ for $s = 1$ and $s = \frac{3}{2}$. The classical ($s \to \infty$) result is also shown.

4. Results

We first show in Fig. 2 our CCM results for the GS magnetic order parameter M. In Fig. 2(a) we consider the spin-1 case, where we show both the "raw" SUBn–n data with $n = 4, 6, 8, 10$ (for the two cases where the CCM model state is chosen to be either the Néel state or the striped state) and the corresponding $n \to \infty$ (SUB∞–∞) extrapolations based on Eq. (15) and using the SUBn–n data set with $n = \{4, 6, 8, 10\}$ as input. One sees that the extrapolated Néel order parameter vanishes at a value $\kappa_{c_1} \approx 0.486$, while the extrapolated striped order parameter similarly vanishes at a value $\kappa_{c_2} \approx 0.527$. As is usually the case, the extrapolations are rather robust with respect to the choice of input data. For example, by comparing similar extrapolations using the alternative data sets with $n = \{4, 6, 8\}$ and $n = \{6, 8, 10\}$, and by making a more detailed error analysis of our results, we find that our best estimates for the QCPs of the spin-1 model are $\kappa_{c_1} = 0.485 \pm 0.005$ and $\kappa_{c_2} = 0.528 \pm 0.005$, as already quoted in Sec. 2. These may be compared with the corresponding best CCM estimates [29] of $\kappa_{c_1} = 0.45 \pm 0.02$ and $\kappa_{c_2} = 0.60 \pm 0.02$ for the QCPs of the spin-$\frac{1}{2}$ model.

We note from Fig. 2(a) that each of the SUBn–n approximants for M terminates at some critical value of κ, beyond which no real solution can be found for the corresponding CCM equations. Such critical values depend both on the model state used and the order n of the SUBn–n approximation. For the Néel state there is an upper such critical value, while for the striped state there is a corresponding lower critical value. These CCM termination points of the coupled sets of SUBn–n equations are typical of the method and are very well understood. Thus, they are simply a reflection of the corresponding QCP that delimits the region of existence for the respective form of magnetic LRO pertaining to that of the CCM model state being used. For any specific finite value n of the SUBn–n truncation index (and for a particular phase under study), Fig. 2(a) clearly demonstrates that the CCM solutions extend into the unphysical regime beyond the actual ($n \to \infty$) QCP out to the respective termination point. As the truncation index n is increased the extent of the corresponding unphysical regime shrinks, and ultimately disappears completely as $n \to \infty$ and the solution becomes exact.

IOP Conf. Series: Journal of Physics: Conf. Series **1041** (2018) 012001 doi:10.1088/1742-6596/1041/1/012001

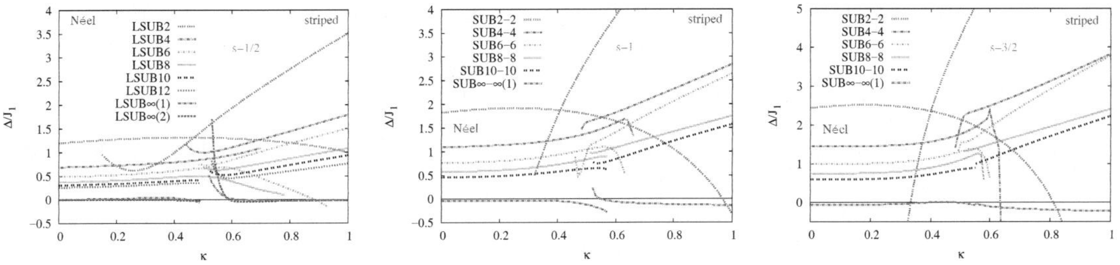

(a) (b) (c)

Figure 3. CCM results for the scaled triplet spin gapΔ $/J_1$ for the J_1–J_2–J_3 model on the honeycomb lattice, with $J_1 > 0$ and $J_3 = J_2 \equiv \kappa J_1 > 0$, as a function of the frustration parameter κ, using both the Néel and striped states as the CCM model state, for the three cases (a) $s = \frac{1}{2}$, (b) $s = 1$, and (c) $s = \frac{3}{2}$. Results are shown in SUBn–n ($\equiv$ LSUBn for $s = \frac{1}{2}$ only) approximations with $n = 2, 4, 6, 8, 10$ in each case and also with $n = 12$ for the case $s = \frac{1}{2}$ only. Extrapolated SUB∞–$\infty(i)$ results are also shown, based in each case on Eq. (16) and the respective data sets $n = \{2, 6, 10\}$ for $i = 1$ and $n = \{4, 8, 12\}$ for $i = 2$ (in the case $s = \frac{1}{2}$ only).

The respective extrapolated CCM results for the scaled magnetic order parameter, M/s, for both the Néel and striped states are compared in Fig. 2(b) for the three cases $s = \frac{1}{2}$, $s = 1$ and $s = \frac{3}{2}$. Results for $s = 2$ and $s = \frac{5}{2}$ have also been given previously in Ref. [33]. One sees clearly that the evidence from our results for the magnetic order parameter alone is that the intermediate phase exists only for the two cases $s = \frac{1}{2}$ and $s = 1$. By contrast, for all cases $s \geq \frac{3}{2}$ the order parameter results indicate a direct transition between the Néel and striped phases at a QCP $\kappa_c(s)$, just as occurs classically in the $s \to \infty$ limit. For the case $s = \frac{3}{2}$ shown in Fig. 2(b), we find $\kappa_c(\frac{3}{2}) = 0.517 \pm 0.004$, based on the order parameter results alone. It has also been observed (and see Ref. [33] for further details) that $\kappa_c(s)$ appears to approach monotonically the classical value $\kappa_{\rm cl} \equiv \kappa_c(\infty) = 0.5$ as s is increased further.

In order to provide further information about the QCPs observed from the magnetic order parameter results in Fig. 2, we now first turn our attention to the triplet spin gap Δ. We display our corresponding CCM results forΔ $/J_1$ as a function of κ in Figs. 3(a), 3(b) and 3(c) respectively for the three cases $s = \frac{1}{2}$, $s = 1$ and $s = \frac{3}{2}$. In each case raw SUBn–n and extrapolated (SUB∞–∞) results are shown based on both the Néel and striped collinear AFM states used separately as our CCM model state. Once again we observe termination points for the excited-state CCM equations, in complete analogy to those discussed above for the corresponding GS equations in connection with the magnetic order parameter results shown in Fig. 2.

At this point we should note that a $(4m - 2)/4m$ staggering effect, where $m \in \mathbb{Z}^+$ is a positive integer, has been observed previously [20, 23, 34, 58] in CCM SUBn–n sequences of results (with n an even integer) for a variety of physical parameters on frustrated honeycomb lattices. Such staggering occurs when the SUBn–n subsequence of results for some parameter with $n = (4m - 2)$ is differentially offset (or staggered) with respect to the corresponding sub-sequence with $n = 4m$. Both sub-sequences still obey an extrapolation scheme of the same sort (i.e., with the same leading exponent), but the coefficients are not identical for both cases. Such an effect is well known both in perturbation theory and in the CCM for all lattices as a corresponding $(2m - 1)/2m$ (i.e., odd/even) staggering. It is precisely for this reason that we do not show SUBn–n results here also for odd values of n, since one should not mix odd and even sub-sequences in a single extrapolation. What is novel for honeycomb-lattice models is an *additional* staggering in the even-order series of terms for some physical observables between those with $n = (4m - 2)$ and those with $n = 4m$. It has been postulated [34] that such an additional staggering can be attributed to the fact that the honeycomb lattice is non-Bravais, comprising two interlocking triangular Bravais lattices, on each of which the usual $(2m-1)/2m$ staggering is observed.

While such a $(4m - 2)/4m$ staggering is not appreciable in the results shown in Fig. 2 for M, it is visibly apparent in the results shown in Fig. 3 for Δ, particularly in the striped phase, but also, albeit to a lesser extent, in the Néel phase. For this reason in our extrapolations forΔ based on Eq. (16) we take care not to mix SUBn–n terms in the input set with $n = (4m - 2)$ and those with $n = 4m$. Thus, for the case $s = \frac{1}{2}$ alone, shown in Fig. 3(a), where we have SUBn–n ($\equiv$ LSUBn in this case) results with $n \leq 12$, we compare the extrapolation based on $n = \{2, 6, 10\}$ with that based on $n = \{4, 8, 12\}$. The agreement between the two is excellent in both stable quasiclassical phases, and again demonstrates the robustness of our extrapolation procedures. The only appreciable difference occurs precisely in the region of the intermediate paramagnetic state near to the boundary with the striped state. Since this is precisely the unphysical part of the region accessible with the striped CCM model state, it is where the associated errors are expected to be the largest and also the most difficult to estimate. What is clear, however, is that the evidence points strongly towards a gapped ($\Delta > 0$) state for the $s = \frac{1}{2}$ case in this intermediate region. For the corresponding cases with $s = 1$ and $s = \frac{3}{2}$ shown in Figs. 3(b) and 3(c) respectively, we are only able to perform SUBn–n calculations for Δ with $n \leq 10$. Hence, for these two cases, we only show the extrapolations for Δ based on Eq. (16) with the data set $n = \{2, 6, 10\}$ used as input.

It is very gratifying to note firstly from Fig. 3 that for both quasiclassical magnetic phases our extrapolated values for Δ are compatible with being zero, within very small numerical errors, over the ranges $\kappa < \kappa_{c_1}$ and $\kappa > \kappa_{c2}$ (with QCPs at κ_{c_1} and κ_{c_2} as determined from Fig. 2 by the points where $M \to 0$) for the two cases $s = \frac{1}{2}$ and $s = 1$, and over the entire range $0 \leq \kappa \leq 1$ for the case $s = \frac{3}{2}$. This is exactly as expected for the case of magnetic LRO, where the low-energy magnon excitations are gapless. The very small negative values shown for parts of the quasiclassical regions in Fig. 3 are clearly just an artefact of the numerical extrapolations. Indeed, their magnitude gives us a good independent check on the overall level of accuracy of our results.

For the case $s = \frac{1}{2}$, shown in Fig. 3(a), there is clear evidence that in the intermediate paramagnetic regime the GS phase is gapped (i.e., with$\Delta > 0$), at least in the regime near κ_{c_2}, where the excited-state LSUBn calculations based on the striped state as CCM model state extend considerably into the "unphysical regime" before the corresponding termination point (discussed previously for the GS solutions) is reached. By contrast, the excited-state LSUBn calculations based on the Néel state terminate for the higher values of n shown, and hence also for the LSUB$\infty(i)$ extrapolants, only just beyond the value κ_{c_1} at which the Néel magnetic order parameter vanishes, as in Fig. 2. Correspondingly, it is more difficult, on the basis of the evidence from Δ alone, to state categorically that the intermediate state is gapped over the entire range $\kappa_{c_1} < \kappa < \kappa_{c_2}$ for the spin-$\frac{1}{2}$ model. Nevertheless, there is clear evidence that at the QCP κ_{c_2} where the striped order melts, the emergent paramagnetic state *is* gapped, and that this gapped state persists over most (if not all) of the regime $\kappa_{c_1} < \kappa < \kappa_{c_2}$.

Turning next to the case $s = \frac{3}{2}$, shown in Fig. 3(c), it is clear that the extrapolant for Δ is zero (within numerical errors) over the entire range of values of the frustration parameter κ shown. For the highest-order SUBn–n calculations shown with $n = 10$, the two results, based separately on the Néel and striped states as CCM model states, terminate at essentially the same point. Furthermore, the two corresponding SUB10-10 results for Δ, shown in Fig. 3(c), lie on a single curve that is continuous and smooth, within very small uncertainties. The same is also true for the SUB∞–$\infty(1)$ extrapolant based on Eq. (16) and the input data set $n = \{2, 6, 10\}$. These results entirely corroborate our findings from Fig. 2 for the order parameter, that there is a direct transition in the spin-$\frac{3}{2}$ case between the two quasiclassical collinear AFM states.

By contrast with the very clear results for Δ for both the $s = \frac{1}{2}$ case shown in Fig. 3(a) and the $s = \frac{3}{2}$ case shown in Fig. 3(c), the results in Fig. 3(b) for the $s = 1$ case are more open to interpretation. For example, in this case, even the two highest-order excited-state SUBn–n solutions shown (i.e., with $n = 10$), based separately on both the Néel and striped states as CCM model state, now show some (possibly unphysical) overlap region where both solutions exist. However, unlike the $s = \frac{3}{2}$ case, the two SUB10-10 solutions in this regime, while close to one another, do not smoothly overlap, as is also the case for the two corresponding SUB∞–∞(1) extrapolants. It seems unlikely that these differences can be attributed to numerical errors, although we cannot wholly rule this out. On balance, the evidence from the results in Fig. 3(b) forΔ , corroborate our finding from Fig. 2(b) that there is a small regime for the spin-1 model between the Néel and striped phases where the stable GS phase is paramagnetic. Furthermore, it also seems reasonably clear that near the QCP where this intermediate state meets the striped state, the former is also gapped, as in the spin-$\frac{1}{2}$ case. From Fig. 3(b) the point where the extrapolant shown for Δ becomes positive, viz., at $\kappa \approx 0.546$ is also very close to the value $\kappa_{c_2} \approx 0.528$ from Fig. 2(b) at which the striped magnetic LRO melts.

Finally, we turn our attention to the two remaining GS low-energy parameters (viz., the spin stiffness coefficient ρ_s and the zero-field transverse magnetic susceptibility χ). Since the results for the cases $s \geq \frac{3}{2}$ are already conclusive (i.e., that there is a direct first-order transition in each case between the Néel and striped AFM phases), we concentrate henceforward on comparing the two cases $s = \frac{1}{2}$ and $s = 1$. Due to the considerably reduced symmetries of both the twisted and canted (Néel and striped) CCM model states required to calculate ρ_s and χ, respectively, we are now only able to perform LSUBn calculations for $s = \frac{1}{2}$ with $n \leq 10$ and SUBn–n calculations for $s \geq 1$ with $n \leq 8$, for both parameters. This can be compared with the corresponding cases $n \leq 12$ and $n \leq 10$ respectively, for the calculations of M and Δ.

In Fig. 4 we display our CCM results for ρ_s (in units of $J_1 d^2$), using both the twisted Néel and twisted canted states separately as model states. The "raw" LSUBn results for the spin-$\frac{1}{2}$ case, shown in Fig. 4(a) are seen to be clearly different in character for the Néel state from their counterparts for the striped state. Thus, firstly, in the former case, the curves flatten and seem to acquire zero slope before or near their termination points, while the latter become steeper near their termination points. This is rather strong evidence for the QCP at κ_{c_1} where Néel order melts being of continuous type, while that at κ_{c_2} where striped order melts is of first-order type. Secondly, whereas the LSUBn results for the Néel state show no perceivable $(4m-2)/4m$ staggering of the sort discussed above, such an effect is clearly seen in their counterparts for the striped state.

This latter difference is also clearly reflected in the behaviour of the extrapolated values. Thus, in Fig. 4(a) we display three different extrapolants LSUB$\infty(i)$, each of which is based on the scheme of Eq. (17), but using the three different input data sets $n = \{4, 6, 8\}$ for $i = 1$, $n = \{6, 8, 10\}$ for $i = 2$ and $n = \{2, 6, 10\}$ for $i = 3$. Clearly, for the Néel phase, the two extrapolants with $i = 1$ and $i = 2$ are nearly identical, thereby demonstrating both the lack of any appreciable $(4m-2)/4m$ staggering and the robustness of our extrapolation procedure using Eq. (17). Interestingly, the only noticeable difference occurs in a very small region near κ_{c_1} where the LSUB∞(2) curve, which utilizes higher-order LSUBn approximants than the LSUB∞(1) curve (and which hence a *priori* is expected to be more accurate), now more closely reflects the continuous nature of the transition with ρ_s approaching zero with zero slope. Even so, the values for κ_{c_1} obtained from the two extrapolations, viz., $\kappa_{c_1} \approx 0.411$ from LSUB∞(1) and $\kappa_{c_1} \approx 0.433$ from LSUB∞(2), are in close agreement. By contrast, the two LSUB$\infty(i)$ extrapolants for the striped phase are not at all in agreement with one another for essentially all values of κ shown. This is clearly a reflection of the now marked $(4m-2)/4m$ staggering, which is clearly perceived by visual inspection of the corresponding LSUBn curves, shown in

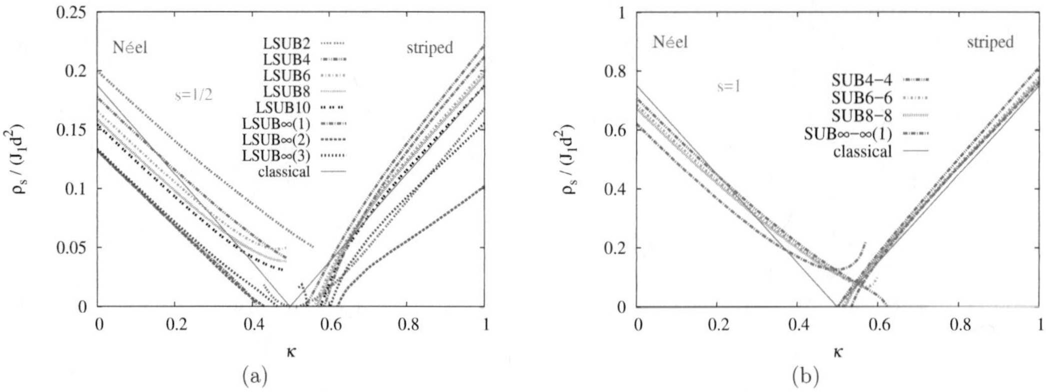

Figure 4. CCM results for the scaled spin stiffness $\rho_s/(J_1 d^2)$ for the J_1–J_2–J_3 model on the honeycomb lattice, with $J_1 > 0$ and $J_3 = J_2 \equiv \kappa J_1 > 0$, as a function of the frustration parameter κ, using both the twisted Néel and twisted striped states as the CCM model state, for the two cases (a) $s = \frac{1}{2}$ and (b) $s = 1$. Results are shown in SUBn–n ($\equiv$ LSUBn for $s = \frac{1}{2}$ only) approximations with $n = 2, 4, 6, 8, 10$ for $s = \frac{1}{2}$ and $n = 4, 6, 8$ for $s = 1$. Extrapolated SUB∞–$\infty(i)$ results are also shown, based in each case on Eq. (17) and the respective data sets $n = \{4, 6, 8\}$ for $i = 1$, $n = \{6, 8, 10\}$ for $i = 2$ and $n = \{2, 6, 10\}$ for $i = 3$ (the latter two in the case $s = \frac{1}{2}$ only). The classical ($s \to \infty$) results from Eqs. (10) and (11) are also shown, using the values $s = \frac{1}{2}$ and $s = 1$ in panels (a) and (b), respectively.

Fig. 4(a), for the striped phase. The corresponding values now obtained for κ_{c_2} (i.e., the points where ρ_s for the striped phase vanishes) are $\kappa_{c_2} \approx 0.588$ from LSUB$\infty(1)$ and $\kappa_{c_2} \approx 0.621$ from LSUB$\infty(2)$.

Obviously, once such $(4m-2)/4m$ staggering effects have been detected, as above, one should base the extrapolations only on LSUBn subsequences with $n = (4m-2)$ alone or with $n = 4m$ alone. For the spin-$\frac{1}{2}$ case, where we have LSUBn results for ρ_s with $n \leq 10$, we can hence use an extrapolation based on Eq. (17) and input data set $n = \{2, 6, 10\}$, which should circumvent any complications of staggering. These are just the LSUB$\infty(3)$ results shown in Fig. 4(a). Again, on the Néel side, all three extrapolations are in good agreement with each other except close to the QCP at κ_{c_1}, whereas on the striped side the LSUB$\infty(3)$ curve should now be clearly taken as our preferred result. The corresponding locations of the two QCPs from the LSUB$\infty(3)$ extrapolation are $\kappa_{c_1} \approx 0.466$ and $\kappa_{c_2} \approx 0.603$. Both are in complete agreement with our previously quoted best CCM estimates [29] of $\kappa_{c_1} \approx 0.45 \pm 0.02$ and $\kappa_{c_2} \approx 0.60 \pm 0.02$.

Turning now to the corresponding spin-1 case, shown in Fig. 4(b), we only have SUBn–n results with $n \leq 8$, and hence we only display the SUB∞–$\infty(1)$ extrapolant based on Eq. (17) and the $n = \{4, 6, 8\}$ input data set. Clearly, on the Néel side the extrapolated value for ρ_s does not vanish, but instead displays a minimum at a value $\kappa \approx 0.495$. This is clearly a shortcoming of the extrapolation, which would presumably disappear if we had higher-order SUBn–n approximants available (e.g., with $n = 10$). Nevertheless, even this limited extrapolant is showing clear evidence for the corresponding QCP being continuous, just as in the spin-$\frac{1}{2}$ case. By contrast, on the striped side, where the QPT is of first-order type, the extrapolated result for ρ_s vanishes at a value $\kappa \approx 0.562$, in reasonable agreement with the previously quoted best CCM estimate [33] of $\kappa_{c_2} \approx 0.528 \pm 0.005$, obtained from the vanishing of the Néel order parameter. In summary, our present results for ρ_s for the spin-1 model reinforce our previous conclusions, while not adding significantly to their accuracy due to the fact that calculations for ρ_s can only be performed to lower orders (for comparable computing resources) than for M.

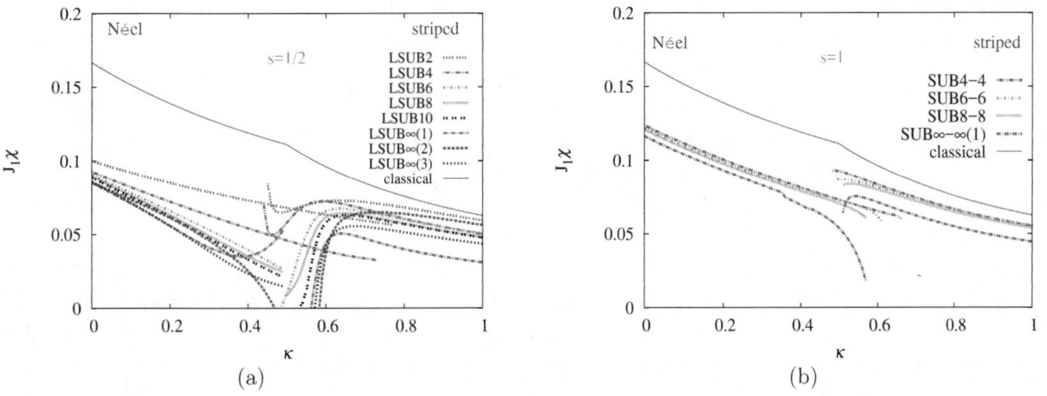

Figure 5. CCM results for the scaled zero-field, transverse magnetic susceptibility $J_1\chi$ for the J_1–J_2–J_3 model on the honeycomb lattice, with $J_1 > 0$ and $J_3 = J_2 \equiv \kappa J_1 > 0$, as a function of the frustration parameter κ, using both the canted Néel and canted striped states as the CCM model state, for the two cases (a) $s = \frac{1}{2}$ and (b) $s = 1$. Results are shown in SUBn–n ($\equiv$ LSUBn for $s = \frac{1}{2}$ only) approximations with $n = 2, 4, 6, 8, 10$ for $s = \frac{1}{2}$ and $n = 4, 6, 8$ for $s = 1$. Extrapolated SUB∞–$\infty(i)$ results are also shown, based in each case on Eq. (18) and the respective data sets $n = \{4, 6, 8\}$ for $i = 1$, $n = \{6, 8, 10\}$ for $i = 2$ and $n = \{2, 6, 10\}$ for $i = 3$ (the latter two in the case $s = \frac{1}{2}$ only). The classical ($s \to \infty$) results from Eqs. (13) and (14) are also shown.

Finally, in Fig. 5, we show corresponding results for the zero-field transverse magnetic susceptibility χ to those shown in Fig. 4 for ρ_s. Two things are immediately apparent from Fig. 5. Firstly, it is evident that the effect of quantum fluctuations is to reduce χ in both quasiclassical AFM phases from its classical value. This reduction is greater for the spin-$\frac{1}{2}$ case than for the spin-1 case, as expected, but is considerable in both. Secondly, and more importantly, we observe from both Figs. 5(a) and 5(b) a marked tendency for χ to vanish at the two QCPs in each case. This is in marked contrast to the classical case where $\chi_{\rm cl}$, also shown in Fig. 5, takes a non-vanishing value ($= \frac{1}{9}$) at the corresponding single classical phase transition at $\kappa_{\rm cl} = \frac{1}{2}$. Thus, in the classical case, when the spins are placed in a transverse magnetic field (of magnitude h) it is always energetically favourable to cant the spins, even in the limit $h \to 0$, and hence $\chi \equiv \chi(h = 0) \neq 0$. By contrast, in the case of a system of quantum spins, if the lowest available excited state has a nonzero excitation energy (i.e.,$\Delta \neq 0$), no excitations are possible (at $T = 0$) as $h \to 0$. Hence, a positive signature of a spin gap opening up at any QCP at which quasiclassical magnetic LRO of any sort melts is the vanishing of χ there [62, 63], indepdendent of the order of the transition.

Turning first to the spin-$\frac{1}{2}$ case, shown in Fig. 5(a), the results for χ show very similar effects (and explanations thereof) to those seen above in Fig. 4(a) for ρ_s. Thus, the shapes of the Néel and striped curves again reflect the different orders of their two transitions into the intermediate phase. Also, on the Néel side, the three extrapolations agree quite well with each other except near the QCP at κ_{c_1}. In particular, only for the LSUB$\infty(2)$ extrapolation, which employs solely the highest-order approximants available (i.e., those with $n = 6, 8, 10$) does χ vanish. The corresponding value $\kappa \approx 0.469$ at which χ vanishes is again in complete agreement with the best CCM estimate for κ_{c_1} (viz., $\kappa_{c_1} \approx 0.45 \pm 0.02$). Similarly to the results for ρ_s in Fig. 4(a), those in Fig. 5(a) for χ in the striped phase also show a marked $(4m-2)/4m$ staggering effect, which is again reflected in the three shown LSUB$\infty(i)$ extrapolants. Corresponding points at which $\chi \to 0$ in the striped phase are, for example, $\kappa \approx 0.583$ for LSUB$\infty(2)$ and

$\kappa \approx 0.572$ for LSUB∞(3), both in reasonable agreement with the best CCM estimate for κ_{c_2} (viz., $\kappa_{c_2} = 0.60 \pm 0.02$). Thus, the results for χ for the spin-$\frac{1}{2}$ case, like those for ρ_s, reinforce our earlier conclusions about its $T = 0$ phase diagram, while again not adding significantly to their accuracy for similar reasons. More importantly, however, they add considerable weight to our finding that the intermediate paramagnetic state is gapped over most (or all) of the region $\kappa_{c_1} < \kappa < \kappa_{c_2}$, particularly at and near the QCP at κ_{c_2}.

Finally, turning to the spin-1 case, shown in Fig. 5(b), we again only have SUBn–n results for values of the truncation index $n \leq 8$, and hence we can only display the SUB∞–∞(1) extrapolant. Again, this has similar shortcomings to those discussed in relation to ρ_s for $s = 1$. While the extrapolated value for χ does not now vanish for either the Néel or striped phase, the tendency to do so is clear. Nevertheless, while these results for χ certainly do not contradict those in Fig. 3(b) for Δ for this case, the evidence for the intermediate state for the spin-1 model to be gapped remains weak at best, except possibly near the QCP at κ_{c_2}.

5. Discussion and conclusions

In this paper we have implemented the CCM to very high orders of approximation in order to investigate the frustrated spin-s J_1–J_2–J_3 Heisenberg antiferromagnet on the honeycomb monolayer lattice. We have concentrated on the case $J_3 = J_2 \equiv \kappa J_1$ and have investigated the $T = 0$ quantum phase diagrams of the model for various values of the spin quantum number s. In particular, we have investigated the phase structure of the model in the most interesting window $0 \leq \kappa \leq 1$ of the frustration parameter κ. This includes the classical tricritical point at $\kappa_{\rm cl} = \frac{1}{2}$ of the full J_1–J_2–J_3 model, which is the point of maximum frustration. The classical model is the limiting case $s \to \infty$, and along the line $J_3 = J_2 \equiv \kappa J_1$ under study the system undergoes a single GS phase transition at $T = 0$, such that for $\kappa < \frac{1}{2}$ the stable phase has AFM Néel LRO, and for $\kappa > \frac{1}{2}$ there exists an IDF of non-planar spin configurations, all degenerate in energy, which includes the collinear state with AFM striped LRO as a special case. Both thermal and leading-order quantum fluctuations at $O(1/s)$ lift the degeneracy in favour of the striped state for $\kappa > \frac{1}{2}$. Our interest here has been to study whether and how quantum fluctuations change the structure of the classical phase diagram for various values of the spin quantum number s.

In order to do so we have first calculated very accurate values for the magnetic order parameter M of both quasiclassical AFM phases for a variety of values of the spin quantum number s. Clear evidence thereby emerged that for the two values $s = \frac{1}{2}$ and $s = 1$ the single classical critical point at $\kappa_{\rm cl} = \frac{1}{2}$ is split into two QCPs, $\kappa_{c_1} < \kappa_{\rm cl}$ and $\kappa_{c_2} > \kappa_{\rm cl}$, such that the system maintains Néel magnetic LRO for $\kappa < \kappa_{c_1}$ and striped magnetic LRO for $\kappa > \kappa_{c_2}$, while the stable GS phase is a paramagnet, with no discernible magnetic LRO, in the intermediate regime $\kappa_{c_1} < \kappa < \kappa_{c_2}$. Our best estimates for the QCPs are $\kappa_{c_1} = 0.45 \pm 0.02$ and $\kappa_{c_2} = 0.60 \pm 0.02$ for the spin-$\frac{1}{2}$ model, and $\kappa_{c_1} = 0.485 \pm 0.005$ and $\kappa_{c_2} = 0.528 \pm 0.005$ for the spin-1 model. By contrast, for all values $s \geq \frac{3}{2}$, the results for M clearly indicate a direct first-order transition between the collinear Néel and striped AFM phases, at a value $\kappa_c(s)$ of the frustration parameter that depends on s and which appears to approach the classical value $\kappa_c(\infty) \equiv \kappa_{\rm cl} = \frac{1}{2}$ monotonically from above. For example, for the spin-$\frac{3}{2}$ model we found $\kappa_c(\frac{3}{2}) \approx 0.517$.

Those findings from the behaviour of $M = M(\kappa)$ were reinforced from further high-order CCM calculations of the triplet spin gap Δ for the three cases $s = \frac{1}{2}, 1, \frac{3}{2}$. In particular, for the spin-$\frac{3}{2}$ model, our results showed that $\Delta = 0$, within very small numerical errors associated with the extrapolations, over the entire range $0 \leq \kappa \leq 1$, compatible with the existence therein everywhere of states with magnetic LRO, whose lowest-lying magnon excitations are gapless. By contrast, for the spin-$\frac{1}{2}$ model, clear evidence emerged that$\Delta \neq 0$ over at least a large part (if not all) of the intermediate regime $\kappa_{c_1} < \kappa < \kappa_{c_2}$ (especially near the second QCP at κ_{c_2}), fully compatible with earlier CCM finding [13] that the paramagnetic state in this case has PVBC order. For the remaining case, $s = 1$, the results from Δ were far less conclusive. Nevertheless, on balance, the evidence does point towards a gapped state opening up at (or very near to) the QCP at κ_{c_2} at which striped order melts.

In view of these results we decided finally to calculate the remaining two low-energy parameters (viz., the spin stiffness ρ_s and the zero-field uniform transverse magnetic susceptibility χ). We decided also to focus on the two cases $s = \frac{1}{2}, 1$ in order to see if we could glean any further evidence from these parameters about the $T = 0$ quantum phase diagram of the spin-1 model, by comparing their behaviours as functions of κ with those of their spin-$\frac{1}{2}$ counterpart. For both cases $s = \frac{1}{2}$ and $s = 1$ the results and conclusions from the calculations of ρ_s largely supported those from the corresponding calculations of M. In particular, from the behaviour of the two curves $M = M(\kappa)$ and $\rho_s = \rho_s(\kappa)$ for both magnetically ordered phases, the evidence is that for both the spin-$\frac{1}{2}$ and the spin-1 model the transition at κ_{c_1} is continuous, while that at κ_{c_2} is first-order.

In principle, calculations of χ can be particularly illuminating since they can provide direct evidence of a gapped state opening, viz., at points where $\chi \to 0$. In the case of the spin-$\frac{1}{2}$ model the evidence was rather compelling that at both QCPs, κ_{c_1} and κ_{c_2}, χ vanishes, thereby reinforcing the belief that the intermediate state in this case is gapped everywhere, compatible with it having PVBC order over the whole intermediate interval $\kappa_{c_1} < \kappa < \kappa_{c_2}$. By contrast, for the spin-1 model the evidence was less conclusive, largely due to the fact that the previously known $(4m - 2)/4m$ staggering for CCM SUBn–n approximants of physical observables for honeycomb-lattice systems is clearly present for χ. Hence, this makes extrapolations of χ particularly problematic for the spin-1 model where SUBn–n approximants to χ are only really practicable (even with large amounts of supercomputing resources available, as here) for values $n \leq 8$ of the truncation parameter, in comparison with $n \leq 10$ for its spin-$\frac{1}{2}$ counterpart.

In summary, the $T = 0$ quantum phase diagrams for the model under study are now well understood from our CCM calculations for all values of the spin quantum number, except $s = 1$. The nature of the intermediate phase for the spin-1 model remains elusive, even after such an exhaustive study as has been performed here. While we have obtained weak evidence that a gapped state opens up as striped order vanishes below κ_{c_2}, this is far from being compelling. Furthermore, there is no direct evidence at all for a gapped state opening above κ_{c_1} where Néel order melts. It would clearly be of great interest to perform SUB10–10 calculations for χ in both magnetic phases for the case $s = 1$, although it seems infeasible that these will become available in the near future. It might also be of interest to use other alternative techniques for the spin-1 model, although they will clearly require very high accuracy to be able to give solid conclusions, on the evidence of the present work.

Acknowledgments
We thank the University of Minnesota Supercomputing Institute for the grant of supercomputing facilities, on which the work reported here was performed. One of us (RFB) gratefully acknowledges Leverhulme Trust for the award of an Emeritus Fellowship (EM-2015-007).

References
[1] Mermin N D and Wagner H 1966 *Phys. Rev. Lett.* **17** 1133–1136
[2] Sachdev S 2008 *Nat. Phys.* **4** 173–185
[3] Sachdev S 2011 *Quantum Phase Transitions* 2nd ed (Cambridge, UK: Cambridge University Press)
[4] Rastelli E, Tassi A and Reatto L 1979 *Physica B & C* **97** 1–24
[5] Mattsson A, Fröjdh P and Einarsson T 1994 *Phys. Rev. B* **49** 3997–4002
[6] Fouet J B, Sindzingre P and Lhuillier C 2001 *Eur. Phys. J. B* **20** 241–254
[7] Mulder A, Ganesh R, Capriotti L and Paramekanti A 2010 *Phys. Rev. B* **81** 214419
[8] Wang F 2010 *Phys. Rev. B* **82** 024419
[9] Cabra D C, Lamas C A and Rosales H D 2011 *Phys. Rev. B* **83** 094506
[10] Ganesh R, Sheng D N, Kim Y J and Paramekanti A 2011 *Phys. Rev. B* **83** 144414

[11] Ganesh R, Sheng D N, Kim Y J and Paramekanti A 2011 *Phys. Rev. B* **83** 219903(E)
[12] Clark B K, Abanin D A and Sondhi S L 2011 *Phys. Rev. Lett.* **107** 087204
[13] Farnell D J J, Bishop R F, Li P H Y, Richter J and Campbell C E 2011 *Phys. Rev. B* **84** 012403
[14] Reuther J, Abanin D A and Thomale R 2011 *Phys. Rev. B* **84** 014417
[15] Albuquerque A F, Schwandt D, Hetényi B, Capponi S, Mambrini M and Läuchli A M 2011 *Phys. Rev. B* **84** 024406
[16] Mosadeq H, Shahbazi F and Jafari S A 2011 *J. Phys.: Condens. Matter* **23** 226006
[17] Oitmaa J and Singh R R P 2011 *Phys. Rev. B* **84** 094424
[18] Mezzacapo F and Boninsegni M 2012 *Phys. Rev. B* **85** 060402(R)
[19] Li P H Y, Bishop R F, Farnell D J J, Richter J and Campbell C E 2012 *Phys. Rev. B* **85** 085115
[20] Bishop R F, Li P H Y, Farnell D J J and Campbell C E 2012 *J. Phys.: Condens. Matter* **24** 236002
[21] Bishop R F and Li P H Y 2012 *Phys. Rev. B* **85** 155135
[22] Li P H Y, Bishop R F, Farnell D J J and Campbell C E 2012 *Phys. Rev. B* **86** 144404
[23] Bishop R F, Li P H Y and Campbell C E 2013 *J. Phys.: Condens. Matter* **25** 306002
[24] Ganesh R, van den Brink J and Nishimoto S 2013 *Phys. Rev. Lett.* **110** 127203
[25] Zhu Z, Huse D A and White S R 2013 *Phys. Rev. Lett.* **110** 127205
[26] Zhang H and Lamas C A 2013 *Phys. Rev. B* **87** 024415
[27] Gong S S, Sheng D N, Motrunich O I and Fisher M P A 2013 *Phys. Rev. B* **88** 165138
[28] Yu X L, Liu D Y, Li P and Zou L J 2014 *Physica E* **59** 41–49
[29] Bishop R F, Li P H Y, Götze O, Richter J and Campbell C E 2015 *Phys. Rev. B* **92** 224434
[30] Zhao H H, Xu C, Chen Q N, Wei Z C, Qin M P, Zhang G M and Xiang T 2012 *Phys. Rev. B* **85** 134416
[31] Gong S S, Zhu W and Sheng D N 2015 *Phys. Rev. B* **92** 195110
[32] Bishop R F and Li P H Y 2016 *J. Magn. Magn. Mater.* **407** 348–357
[33] Li P H Y, Bishop R F and Campbell C E 2016 *J. Phys.: Conf. Ser.* **702** 012001
[34] Li P H Y and Bishop R F 2016 *Phys. Rev. B* **93** 214438
[35] Bishop R F and Kümmel H G 1987 *Phys. Today* **40(3)** 52–60
[36] Bishop R F 1991 *Theor. Chim. Acta* **80** 95 148
[37] Bishop R F 1998 *Microscopic Quantum Many-Body Theories and Their Applications* Lecture Notes in Physics Vol. 510 ed Navarro J and Polls A (Berlin: Springer-Verlag) pp 1–70
[38] Bartlett R J and Musiał M 2007 *Rev. Mod. Phys.* **79** 291–352
[39] Coester F 1958 *Nucl. Phys.* **7** 421–424
[40] Coester F and Kümmel H 1960 *Nucl. Phys.* **17** 477–485
[41] Čižek J 1966 *J. Chem. Phys.* **45** 4256–4266
[42] Čižek J 1969 *Adv. Chem. Phys.* **14** 35–89
[43] Bishop R F and Lührmann K H 1978 *Phys. Rev. B* **17** 3757–3780
[44] Kümmel H, Lührmann K H and Zabolitzky J G 1978 *Phys Rep.* **36** 1–63
[45] Bishop R F and Lührmann K H 1982 *Phys. Rev. B* **26** 5523–5557
[46] Arponen J 1983 *Ann. Phys. (N.Y.)* **151** 311–382
[47] Arponen J S, Bishop R F and Pajanne E 1987 *Phys. Rev. A* **36** 2519–2538
[48] Arponen J S, Bishop R F and Pajanne E 1987 *Phys. Rev. A* **36** 2539–2549
[49] Arponen J S and Bishop R F 1991 *Ann. Phys. (N.Y.)* **207** 171–217
[50] Stanton J F and Bartlett R J 1993 *J. Chem. Phys.* **98** 7029–7039
[51] Arponen J S and Bishop R F 1993 *Ann. Phys. (N.Y.)* **227** 275–333
[52] Arponen J S and Bishop R F 1993 *Ann. Phys. (N.Y.)* **227** 334–380
[53] Zeng C, Farnell D J J and Bishop R F 1998 *J. Stat. Phys.* **90** 327–361
[54] Farnell D J J and Bishop R F 2004 *Quantum Magnetism* Lecture Notes in Physics Vol. 645 ed Schollwöck U, Richter J, Farnell D J J and Bishop R F (Berlin: Springer-Verlag) pp 307–348
[55] Bishop R F, Li P H Y and Campbell C E 2014 *AIP Conf. Proc.* **1619** 40–50
[56] Bishop R F, Li P H Y, Zinke R, Darradi R, Richter J, Farnell D J J and Schulenburg J 2017 *J. Magn. Magn. Mater.* **428** 178–188
[57] Villain J 1977 *J. Phys. (France)* **38** 385–391
[58] Bishop R F and Li P H Y 2017 *Phys. Rev. B* **96** 224416
[59] Senthil T, Vishwanath A, Balents L, Sachdev S and Fisher M P A 2004 *Science* **303** 1490–1494
[60] Senthil T, Balents L, Sachdev S, Vishwanath A and Fisher M P A 2004 *Phys. Rev. B* **70** 144407
[61] We use the program package CCCM of D. J. J. Farnell and J. Schulenburg, see http://www-e.uni-magdeburg.de/jschulen/ccm/index.html
[62] Mila F 2000 *Eur. J. Phys.* **21** 499–510
[63] Bernu B and Lhuillier C 2015 *Phys. Rev. Lett.* **114** 057201

IOP Conf. Series: Journal of Physics: Conf. Series **1041** (2018) 012002 doi:10.1088/1742-6596/1041/1/012002

Metal-insulator transition in the hybridized two-orbital Hubbard model revisited

Y. Núñez-Fernández and K. Hallberg

Centro Atómico Bariloche and Instituto Balseiro, CNEA and CONICET, 8400 Bariloche, Argentina

E-mail: yurielnf@gmail.com

Abstract. In this work we study the two-orbital Hubbard model on a square lattice in the presence of hybridization between nearest-neighbor orbitals and a crystal-field splitting. We use a highly reliable numerical technique based on the density matrix renormalization group to solve the dynamical mean field theory self-consistent impurity problem. We find that the orbital mixing always leads to a finite local density states at the Fermi energy in both orbitals when at least one band is metallic. When one band is doped, and the chemical potential lies between the Hubbard bands in the other band, the coherent quasiparticle peak in this orbital has an exponential behavior with the Hubbard interaction U.

1. Introduction

The Mott-Hubbard metal-insulator transition (MIT) remains a highly interesting problem in correlated systems due to its relevance in many families of materials interesting physical properties and with current and potential applications. [1, 2] Most interesting materials such as transition metal oxides, which involve more than one correlated band, lack appropriate theoretical studies due to the complexity of the corresponding models. Physical considerations of crystal field effects and the Coulomb exchange energy are relevant for the atomic energies. In addition, a non-local inter-orbital hopping should be considered in more realistic models since it is also generally present and it can vary significantly for different orbital components and orientations.

An interesting feature in these materials is whether an orbital-selective Mott (OSM) localization occurs, in which some electrons remain itinerant and others undergo a Mott transition. This is thought to happen for example, for some Lanthanides and Actinides (with localized electrons in incomplete f-shells and itinerant s, p and d electrons), in manganites like $La_{1-x}Sr_xMnO_3$ (with localized electrons in the t_{2g} orbital of Mn and delocalized e_g electrons)[3], and also for $Ca_xSr_{2-x}RuO_4$ and LiV_2O_4 for which selective localization can occur for electrons in the same shell.[4]

Among the most relevant numerical methods to treat these systems, the dynamical mean field theory (DMFT) [5] has triggered a great deal of theoretical research leading to important progress in the understanding of the electronic structure in these systems.

At the core of the DMFT method is the calculation of Greens functions of an interacting quantum impurity for which the fermionic bath has to be determined self-consistently until convergence of the local Green function and the local self-energy is reached. Recent advances in numerical techniques for the resolution of multiorbital quantum impurity problems have made possible the implementation of the Dynamical Mean Field Theory (DMFT) in model Hamiltonians and Density-Functional-based methods for the calculation of strongly correlated materials in a realistic way.[6]

One of the most reliable impurity solvers is the Density Matrix Renormalization Group (DMRG)[7, 8, 9, 10, 11, 12] which allows for the calculation of the density of states directly on the real axis and at zero temperature (or with a very small imaginary offset). This methods complements Monte Carlo based impurity solvers such as the CTQMC [13, 14] which suffer from the fermionic sign problem (particularly for hybridized models), from approximate analytic continuation of the Green's functions to the real energy axis and are useful for finite temperatures. It is comparable, though, to the NRG-based impurity solver.[15, 16, 17]

In this paper we use the DMFT with the DMRG as the impurity solver to study the hybridized two-orbital Hubbard model, which captures the main physics of many correlated systems; in particular we are interested in the effect of the inter-orbital hopping and the crystal-field splitting in the Mott transition.

2. Model and method

We consider the following model:

$$H = \sum_{ij\alpha\beta} t_{ij}^{\alpha\beta} c_{i\alpha}^{\dagger} c_{j\beta} - \mu \sum_{i} n_{i}, + U \sum_{\alpha} n_{i\alpha\uparrow} n_{i\alpha\downarrow} \tag{1}$$

where $c^{\dagger}$ creates an electron, i, j are sites indices on the square lattice, $\alpha,\beta = 1, 2$ are the orbital indices, spin indices are implicit, and we consider the paramagnetic solution. Here we define the on-site and hopping operators in the orbital basis:

$$\mathbf{t}_0 = \begin{pmatrix} \Delta & 0 \\ 0 & -\Delta \end{pmatrix}, \; \mathbf{t}_x = \begin{pmatrix} t & t' \\ t' & t \end{pmatrix}, \; \mathbf{t}_y = \begin{pmatrix} t & -t' \\ -t' & t \end{pmatrix}, \tag{2}$$

where the subindices indicate the on-site, x and y hopping matrices, Δ is the crystal field splitting and t' is the inter-orbital hopping parameter. We will consider $t = 0.25$ as the unit of energy.

Defining the matrix $T(k)$ in the reciprocal space $T(k) = \mathbf{t}_0 + 2\mathbf{t}_x \cos k_x + 2\mathbf{t}_y \cos k_y - \mu\mathbf{I}$ and the self-energy $\Sigma(k, z)$, $z = \omega + i\eta$ we can write the Green's function as:

$$G(k, z) = [z\,I - T(k) - \Sigma(k, z)]^{-1} \tag{3}$$

It can be shown that, in the half-filled case, the Hamiltonian (1) is invariant under the transformation $c_{i\alpha}^{\dagger} \to (-1)^{x_i+y_i+\alpha} c_{i\bar{\alpha}}$ and that the Green's function is diagonal due to the reflection symmetry σ_d in the square lattice, $G_{12}(\omega) = G_{21}(\omega) = 0$:

$$G(z) = \begin{pmatrix} G_{11}(z) & 0 \\ 0 & G_{22}(z) \end{pmatrix}, \tag{4}$$

where $G_{22}(z) = G_{11}(-z)$ for half filling. This implies that the self-energy (Σ) and hybridization (Γ) matrices are also diagonal (see below).

We use the DMFT to obtain the electronic densities of states in this model. This technique approximates the momentum dependent self-energy by a local quantity $\Sigma(k, z) \approx \Sigma(z)$. The hybridization used in this method has the same structure as (4) and this means that both orbitals are decoupled, having independent hybridization baths, which simplifies the calculation. We define the effective impurity Hamiltonian H_{imp} taking part in the DMFT equations as:

$$H_{imp} = \sum_{\alpha} t_0^{\alpha\alpha} c_{0\alpha}^\dagger c_{0\alpha} + U \sum_{\alpha} n_{0\alpha\uparrow} n_{0\alpha\downarrow} + H_b, \tag{5}$$

where the non-interacting bath term H_b is:

$$H_b = \sum_{\alpha i} \lambda_i^\alpha b_{\alpha i}^\dagger b_{\alpha i} + \sum_{\alpha i} v_i^\alpha \left[b_{\alpha i}^\dagger c_{0\alpha} + H.c. \right], \tag{6}$$

where $b_{\alpha i}^\dagger$ represents the creation operator for the site i associated to the bath site α.

The DMFT iterations are the following:

(i) Define $\Sigma(z) = 0$,

(ii) Calculate the Green's function at the impurity site for each orbital and the hybridization:

$$G_{\alpha\alpha}(z) = \frac{1}{N} \sum_k G_{\alpha\alpha}(k, z) = \frac{1}{N} \sum_k \left[z\, I - T(k) - \Sigma(z) \right]_{\alpha\alpha}^{-1} \tag{7}$$

$$\Gamma_{\alpha\alpha}(z) = z - t_0^{\alpha\alpha} - \Sigma_{\alpha\alpha}(z) - G_{\alpha\alpha}^{-1}(z). \tag{8}$$

(iii) Find an approximation $\tilde{\Gamma}_{\alpha\alpha}(z)$ for the hybridization$\Gamma\ _{\alpha\alpha}(z)$ via the parameters v_i^α and λ_i^α:

$$\tilde{\Gamma}_{\alpha\alpha}(z) = \sum_i \frac{|v_i^\alpha|^2}{\omega - \lambda_i^\alpha}. \tag{9}$$

(iv) Calculate the Green's function $G_{\alpha\alpha}(z)$ at site "0" of H_{imp} using DMRG:

(v) Calculate the self-energy:

$$\Sigma_{\alpha\alpha}(z) = z - t_0^{\alpha\alpha} - G_{\alpha\alpha}^{-1}(z) - \tilde{\Gamma}_{\alpha\alpha}(z). \tag{10}$$

Return to **(ii)** until convergence.

3. Results

Using the method described above, we obtained the densities of state (DOS) of model (1) for several representative parameters. In Fig. 1 we show the behavior of both orbitals for different values of the interaction U and a finite crystal field splitting Δ. Also shown is the particle filling of each orbital vs. U. We find that, due to the interorbital hopping t', the lower-lying orbital (orbital 2) develops a large weight at the Fermi energy. Both orbitals transition to an insulating state upon increasing U as seen also in the filling values, approaching the half-filled situation ($n = 1$) simultaneously.

For small interactions we find a non-correlated metal-band insulator transition by varying the crystal-field splitting, as shown in Fig. 2.

In Fig. 3 we vary the chemical potential and see that, for these parameters, orbital 2 remains half-filled while band 1 empties by increasing the absolute value of μ. A finite DOS at the Fermi energy in orbital 1 always leads to a finite DOS in orbital 2, seen here as a small peak at zero energy.

It is interesting to analyze the effect on orbital 2 of a large density of states in orbital 1. We find that, when the Fermi energy lies in the lower Hubbard band of the upper orbital (orbital 1), it causes a finite DOS in the lower orbital at the Fermi energy, even for large values of U, see Fig. 4. As a consequence, orbital 2 does not have a Mott metal-insulator transition and there is no selective MIT.[18]

IOP Conf. Series: Journal of Physics: Conf. Series **1041** (2018) 012002 doi:10.1088/1742-6596/1041/1/012002

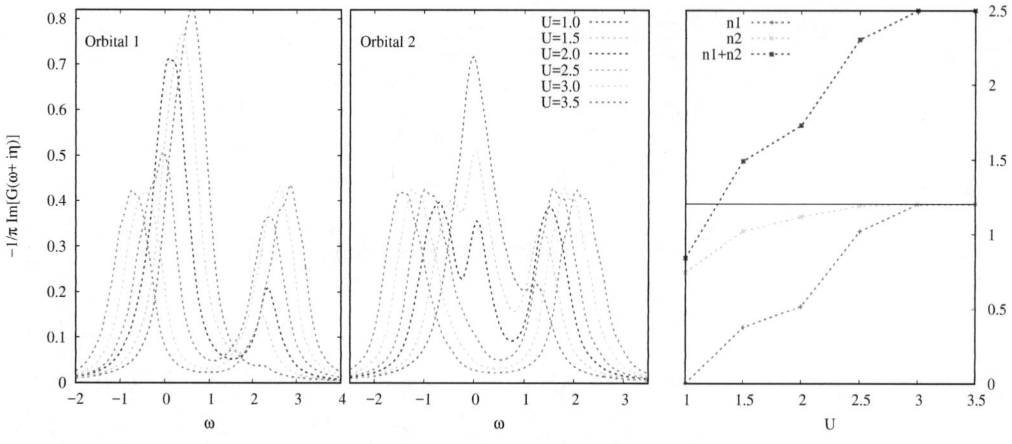

Figure 1. DOS for the hybridized two-band Hubbard model with a finite crystal-field splitting $\Delta = 0.3$, $t' = 0.15$ and $\mu = -0.7$, increasing U showing a simultaneous metal-insulator transition in orbitals 1 and 2. Note that the finite DOS at the Fermi energy ($\omega = 0$) for the insulating state for $U \geq 3$ is due to the finite η Lorentzian broadening (see text before Eq. 3). Right: Orbital fillings vs. U.

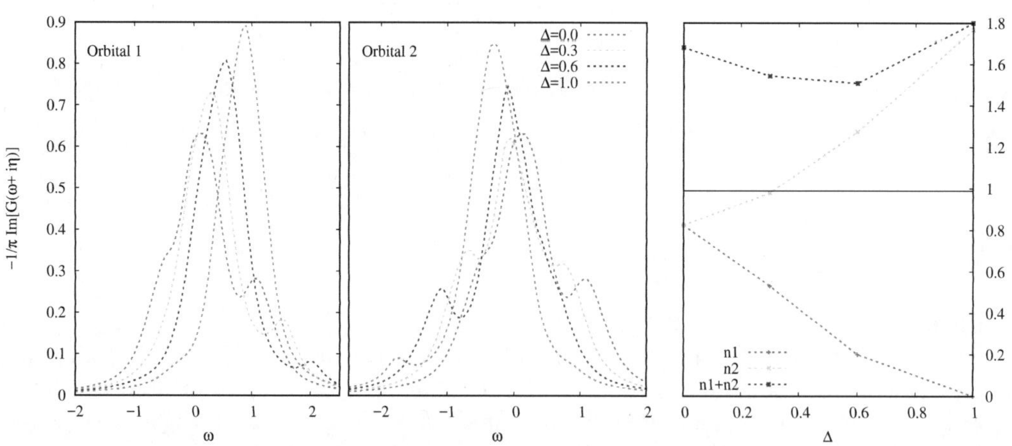

Figure 2. DOS for $U = 1$, $t' = 0.15$ and $\mu = -0.3$, changing Δ showing transfer of electronic filling from orbital 1 to 2. Right: The corresponding orbital fillings.

We measured the integral of the peak at the Fermi energy of orbital 2 (which is proprtional to the quasiparticle weight Z_2). We find that, as long as the Fermi energy lies within the Hubbard band in orbital 1, the quasiparticle weight in orbital 2 will be finite. All other parameters fixed, Z_2 decreases exponentially with U like in the Kondo problem (see Fig. 3), not showing a metal-insulator transition for any value of U. The larger the inter-orbital hopping, the larger the DOS at the Fermi energy in orbital 2 (see Fig. 4).

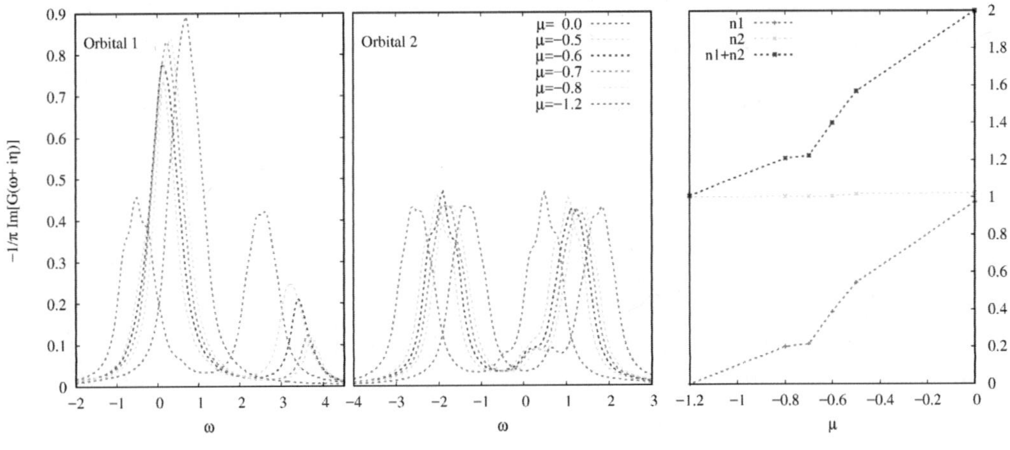

Figure 3. Same as Fig. 2 but for $U = 3$, $\Delta = 1$, for different chemical potentials and their corresponding fillings (right).

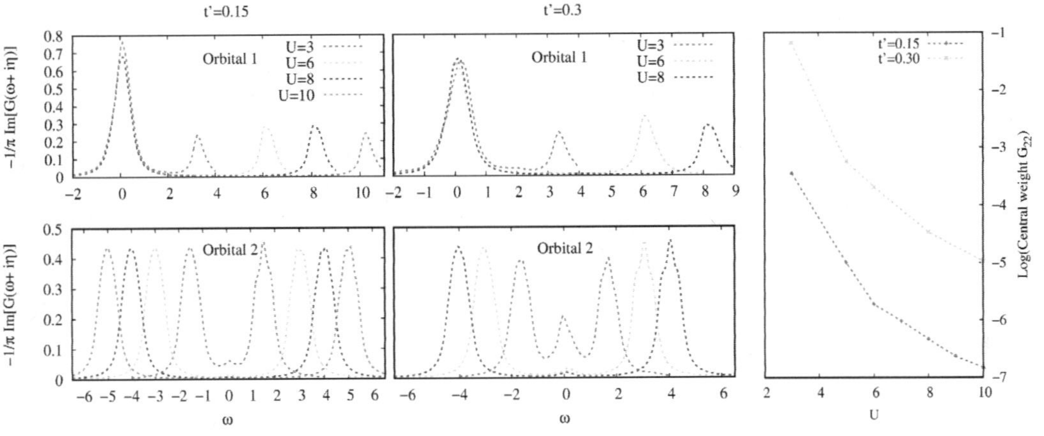

Figure 4. DOS for fixed $\Delta = 0.75$ and $\mu = -0.75$ for which the Fermi energy lies in the lower Hubbard band for orbital 1 and between both Hubbard bands in orbital 2 for two values of inter-orbital hybridization t'. Note the shift in energy values between the left and middle panels. Right: Quasiparticle weight in orbital 2 vs. U.

4. Conclusions

By using a reliable zero temperature technique which is free from fermionic signs and delivers the Green functions on the real energy axis directly to solve the DMFT impurity problem, we calculate the density of states of the hibridized two-band Hubbard model on a square lattice. We find that for finite inter-orbital mixing, there is a finite density of states at the Fermi energy of both bands even for large interactions, as long as one band remains metallic. The metal-insulator transition, if it occurs, happens simultaneously for both orbitals. By doping one of the bands and fixing the chemical potential in between the Hubbard bands in the other correlated orbital we find that the density of states in this second orbital remains finite even for large interactions. Here we do not find any MIT and the correlated metallic quasiparticle peak decreases exponentially with with the Hubbard interaction U.

IOP Conf. Series: Journal of Physics: Conf. Series **1041** (2018) 012002 doi:10.1088/1742-6596/1041/1/012002

4.1. Acknowledgements
We acknowledge support from projects PICT 2012-1069 and PICT 2016-0402 from the Argentine ANPCyT and PIP 2015-2017 11220150100538CO (CONICET). This work used the Extreme Science and Engineering Discovery Environment (XSEDE), which is supported by National Science Foundation grant number ACI-1548562.

References
[1] N. Mott, Metal-Insulator Transitions (Taylor and Francis Ltd., London/Philadelphia, 1990).
[2] M. Imada, A. Fujimori, and Y. Tokura, Rev. Mod. Phys. **70**, 1039 (1998).
[3] M. B. Salomon and M. Jaime, Rev. Mod. Phys. 73, 583 (2001)
[4] L de Medici, G Giovannetti, M Capone, Phys. Rev. Lett. 112, 177001 (2014)
[5] A. Georges, G. Kotliar, W. Krauth and M. J. Rozenberg, Rev. Mod. Phys. **68**, 13 (1996)
[6] K. Hallberg, D. García, Pablo S. Cornaglia, Jorge I. Facio and Y. Núñez Fernández, EPL Perspectives **112** 17001 (2015)
[7] S. White, Phys. Rev. Lett. **69**, 2863 (1992); S. White, Phys. Rev. B **48**, 10345 (1993)
[8] U. Schollwöck, Rev. Mod. Phys. **77**, 259 (2005)
[9] K. Hallberg, Adv. Phys. **55**, 477-526 (2006)
[10] D. J. García, K. Hallberg, and M. J. Rozenberg, Phys. Rev. Lett. **93**, 246403 (2004)
[11] D. J. García, E. Miranda, K. Hallberg, M. J. Rozenberg, Phys. Rev. B **75**, 121102(R) (2007); E.Miranda, D. J. García, K.Hallberg, and M. J. Rozenberg, Physica B: Cond. Mat. **403**, 1465 (2008); D. J. García, E.Miranda, K.Hallberg, and M. J. Rozenberg, Physica B: Cond. Mat. **398**, 407 (2007)
[12] Y. Núñez Fernández, D. García, and K. Hallberg, Journal of Physics: Conference Series **568**, 042009 (2014)
[13] A. N. Rubtsov, V. V. Savkin, and A. I. Lichtenstein, Phys. Rev. Lett. **72**, 035122 (2005)
[14] E. Gull, A. J. Millis, A. I. Lichtenstein, A. N. Rubtsov, M. Troyer, and P. Werner, Rev. of Mod. Phys. **83**, 349 (2011)
[15] R. Bulla, Phys. Rev. Lett. **83**, 136 (1999); R. Bulla, A. C. Hewson, and T. Pruschke, J. Phys.:Cond. Mat. **10**, 8365 (1998)
[16] R. Bulla, T. Costi and D. Vollhardt, Phys. Rev. B **64**, 45103 (2001)
[17] R. Zitko and T. Pruschke, Phys. Rev. B **79**, 085106 (2009)
[18] L. de Medici, A. Georges, G. Kotliar, and S. Biermann, Phys. Rev. Lett. **95**, 066402 (2005)

IOP Conf. Series: Journal of Physics: Conf. Series **1041** (2018) 012003 doi:10.1088/1742-6596/1041/1/012003

Physics of Narrow Superconducting and Superfluid Channels: Critical Temperature and Associated Single Vortices Nucleation

M B Sobnack and F V Kusmartsev
Department of Physics, Loughborough University, Loughborough LE11 3TU, United Kingdom
E-mail: m.b.sobnack@lboro.ac.uk

Abstract. It is well known that in lower dimensions at any finite temperature one can thermally create single vortex excitations. This is because the boundary cut-off eliminates the logarithmic divergence of the vortex energy, thus eliminating the Berezinskǐ-Kosterlitz-Thouless transition which typically exists in 2D films. However, single vortices created in a channel are strongly attracted to the boundaries; as a result, their life-time is very short. Here we demonstrate that with rising temperature the density of spontaneously generated vortices and anti-vortices increases significantly faster than one would expect from conventional activation law. The two boundaries of the channels act as mirrors that multiply the effective number of vortices and anti-vortices spontaneously created and makes it practically equal to infinity. All these vortices, a few real bare and infinitely many imaginary ones, screen the vortex-vortex and vortex-boundary interactions. As a result of the screening, the life-time and density of vortices increase. This process constitutes a new mechanism for the destruction of phase coherence, superconductivity or superfluidity in narrow channels. We describe this effect using real space renormalization group techniques and derive an expression for the scale-dependent critical temperature at which the vortices destroy superconductivity/superfluidity in a cascade manner.

1. Introduction

High-temperature superconductors (HTS) are most promising materials for numerous practical applications. These include HTS magnets, and SQUIDs. Indeed, over the past years, there has been an enormous interest in the superconducting properties of narrow channels and quasi-one-dimensional systems [1, 2, 3, 4, 5, 6, 7, 8]. In particular, there is renewed interest in the breakdown of superconductivity in systems with confined geometries [8]. It has been shown that the geometry and the size determine the dynamics of the vortices in the presence of the transport current subject to a magnetic field orthogonal to the axis [9]. Moreover, it was shown that the character of the superconducting phase slips depends on the transverse size of the superconducting nanowires. For ultra-narrow nanowires, the phase slips ascribe a quantum character that change the character of the superconducting state and can even completely suppress it [8]. In this manuscript, we study the suppression of superconductivity in narrow superconducting thin tapes and nanowires and, in general, in narrow superfluid channels. However, in contrast with the above studies [8, 9] we do not study ultra narrow superconducting wires where quantum phase slips may occur, but we study superconducting/supefluid channels with widths much larger than the superconducting coherence length. We show that there is another mechanism of the suppression of the superconducting state which arises at high temperatures and is associated with the spontaneous cascade creation of vortices from the boundary.

The essence of the new mechanism is in the influence of the boundary on the vortex nucleation. For the narrow superconducting wires, the low energy excitations of the systems are the vortices created near the boundary. The point is that the energy required to create a quantised vortex in a system of characteristic size L depends logarithmically on the size of the system, *i.e.*, it requires an enormous amount of energy to create the stationary vortex. This is the main reason why these topological defects did not receive much attention when they were proposed by Onsager [10] in the 40's and by Feynman [11] in the 50's, and observed by Vinen [12] in the 70's. It was not until Berezinskǐ [13] showed that it costs much less energy to create a vortex- anti-vortex (V–A) pair (with the energy depending logarithmically on the pair separation) that their importance was realised. Berezinskǐ recognised that these defects may play a significant role in phase transitions and qualitatively described the phase transitions stimulated by spontaneous vortex- anti-vortex pair generation. Kosterlitz and Thouless [14, 15] described the scaling relations of the transition. Since then, phase transitions in two dimensions (2D) have been the subject of a lot of interest given that the Berezinskǐ -Kosterlitz-Thouless (BKT) transition [13-15] [13, 14, 15] must arise in 2D systems such as superfluid films and quasi-2D superconductors.

Qualitatively, the BKT transition involves the nucleation of V–A pairs, which reduces the superfluid density. This, in turn, diminishes the binding energy of V–A pairs so that it becomes easier for more pairs to be created, reducing the superfluid density further. This causes a chain (renormalization) process that continues until the screening effect is large enough for members of V–A pairs to become completely unbound. Vortices and anti-vortices then nucleate freely and spontaneously. Since single vortices destroy the phase coherence needed for 2D superfluidity, the superfluidity is destroyed and there is a jump discontinuity in the superfluid density at the transition.

The situation may be rather different in the case of "small" superfluid samples or in a near-edge layer (or a stripe) of macroscopic films. Due to the confinement, there is a cut-off in the interaction energy for individual vortices — not only for V–A pairs — while the condensation energy associated with the vortex cores is twice smaller for the case of single vortex excitations. This can make nucleation of single vortices in confined geometries energetically more favorable than nucleation of V–A pairs. In this study, we show that in this case, a single–vortex plasma arising due to thermal fluctuations can destroy superfluidity, playing the same role as the V–A plasma does in the BKT transition existing in two-dimensional superfluid and superconducting films. Note that, for narrow superconducting films, the interaction between vortices and anti-vortices is screened due to the existence of the hard boundary. In such a strip, each hard boundary plays the role of a mirror and therefore each vortex or anti-vortex created in the strip is attracted to an infinite number of image vortices and anti-vortices created by the mirror. As a result, the vortex charge is spread over the line of its images and the images of the images, therewith leading to a complete screening of the vortex-vortex interaction. If such a screening would arise in a superconducting film, the superconducting state would not exist. The existence of the superconducting state in the narrow strip is due to the boundary which attracts the vortices or anti-vortices and therewith removing them from the strip, cleaning the coherent state. However, the attraction of vortices to the boundaries can be also screened and the effect depends on the width of the superconducting strip and increases when the temperature rises.

Thus, in our approach, we describe the interaction of a single vortex with two sample edges, which can effectively be described as its attraction to infinitely many imaginary vortices (IV) and anti-vortex (IA), located beyond the sample (vortices, just like 2D electric charges, are attracted to their mirror anti-charges). A vortex can penetrate into the sample only by overcoming

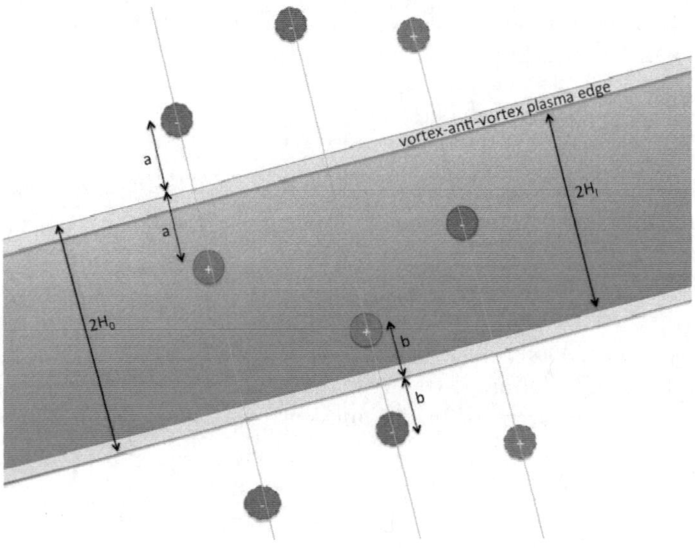

Figure 1. This figure shows schematically the channel (of width $2H = 2H_0$) studied in the paper. A vortex (or anti-vortex) is strongly attracted to the boundary, and the energy required to create such a vortex (anti-vortex) is calculated, in analogy with 2D electrostatics, by calculating the energy of interaction of the vortex (anti-vortex) with an infinite number of image vortices and an infinite number of image anti-vortices situated beyond the channel, giving a logarithmic energy. Even one created vortex has an infinitely many vortex and antivortex images. There even a few created vortices act and move as in a plasma of vortices. As the temperature increases, there will be further vortex (anti-vortex) excitations in the system, including some in the space between the vortex (anti-vortex) and the boundary. This vortex (anti-vortex) plasma edge renormalizes the attraction of the vortex (anti-vortex) to the boundary. The width of this plasma edge increases when the temperature rises and becomes equal to the width of the channel at the critical temperature. The problem is studied here using renormalization group techniques.

this vortex attraction to their multiple images of vortices and anti-vortices (V–IA and V–IV) interaction, which has a logarithmic character. The same is true for an anti-vortex, which can penetrate into the sample only by overcoming a similar anti-vortex to multiple images of vortices (A–IV) and anti-vortices (A–IA) interaction. For bulk superconductors, this attraction to the boundary is known as a surface or Bean-Livingston barrier. Let us assume that several thermally excited vortices have been created near the boundary. These non-stationary vortices induce infinitely many images which renormalize (decrease) the vortex attraction to the boundary. In a fashion similar to the screening of the V–A interaction in the BKT mechanism, now the imaginary plasma of V–IA , V–IV screens the "Coulomb" attraction. Eventually this chain process leads to the nucleation of unbound fluctuational vortices and anti-vortices in a relatively wide layer near the boundary and the order parameter associated with superfluidity is destroyed. The temperature at which this crossover happens can be defined as a new size–dependent critical temperature which is lower than the critical temperature T_c of the bulk material. Because this crossover occurs very fast, the described phenomenon presents the same kind of topological phase transition as the BKT [13-15] [13, 14, 15] transition, the destruction of superconductivity in a superconducting disk [16] or the Williams-Shenoy model for the λ–phase transition in superfluid Helium [17, 18]. Note that the screening process described here in which one vortex generates infinitely many image vortices and antivortices has never been discussed before. This mechanism

IOP Conf. Series: Journal of Physics: Conf. Series **1041** (2018) 012003 doi:10.1088/1742-6596/1041/1/012003

is completely different from the one which we described in our previous PRL paper [16] where each vortex has only one (not infinitely many, as in the present case) image anti-vortex taking part in the vortex dynamics.

2. Formulation

2.1. Vortices in superconducting tapes and analogy with flows in classical hydrodynamics

The model we study is a quasi-1D channel of superfluid film (density ρ_s) of width $2H$, lying between $x = -H$ and $x = H$ and unbounded in the y-direction. Consider a vortex of vorticity $\kappa = h/m$ at an arbitrary position $\mathbf{r}_0 = (x_0, y_0)$, $0 < x_0 < H$. The correct boundary condition on $x = \pm H$ is that any flow at the boundaries should be tangential, *i.e.*, the normal component of the flow at the boundaries should be zero. This is achieved, in analogy with 2D electrostatics, by having an infinite array of image vortices (each of vorticity κ) beyond the channel at $\mathbf{r}_m = (x_m, y_0)$ and an infinite array of image anti-vortices (of vorticity $-\kappa$) at $\mathbf{r}'_m = (x'_m, y_0)$, where

$$x_m = x_0 + 4Hm, \quad m \in \mathbb{Z}, m \neq 0$$

and

$$\begin{aligned} x'_m &= 2H - x_0 + 4Hm, \quad m \in \mathbb{Z} \\ &= x'_0 + 4Hm. \end{aligned}$$

The energy required to create the vortex at $\mathbf{r}_0$ is the energy with which it is attracted to the boundaries. This can be evaluated by calculating the energy of interaction of the vortex with all the image vortices and anti-vortices. We do this in analogy with 2D hydrodynamics (see, for example, Newton [19] and references therein) and after a lengthy calculation we arrive at

$$U_0(x_0) = 2q^2 \ln \frac{2\cos kx_0}{kr_c} + E_c \tag{1}$$

for the energy required to create a vortex in the channel at $\mathbf{r_0} = (x_0, y_0)$. r_c is the radius of the core of the vortex, E_c is the potential energy associated with the vortex core and, where in analogy with the 2D Coulomb gas,

$$q = \frac{1}{2}\kappa \left(\frac{\rho_s}{\pi}\right)^{1/2} = \frac{\hbar}{m}\sqrt{\pi \rho_s}$$

is the effective vortex charge. The core energy E_c is sometimes written as $-\mu$ and is called the chemical potential of the vortex. $U_0(x_0)$ is the energy with which the vortex charge at $\mathbf{r_0}$ is attracted to the boundary (surface) of the superfluid material. From now on, we shall drop the subscript "0" from x_0.

2.2. Screening of the vortex attraction to the boundaries and the renormalization group equations

At low temperatures, it is likely that only a few vortices will be present. These cannot nucleate and are attracted to the boundary. This will also happen with virtual vortices and anti-vortices originated from a vacuum state as well. At higher temperatures, however, there are likely to be many more vortex excitations, including some located in the space between x and H. These have an attenuating effect on, and screen, the interaction $U_0(x)$. To take into account this screening effect, we follow Kosterlitz and Thouless [14, 15], Sobnack [16], Williams [17] and Shenoy [18] by introducing a scale-dependent dielectric constant

$$\varepsilon = 1 + 4\pi\chi. \tag{2}$$

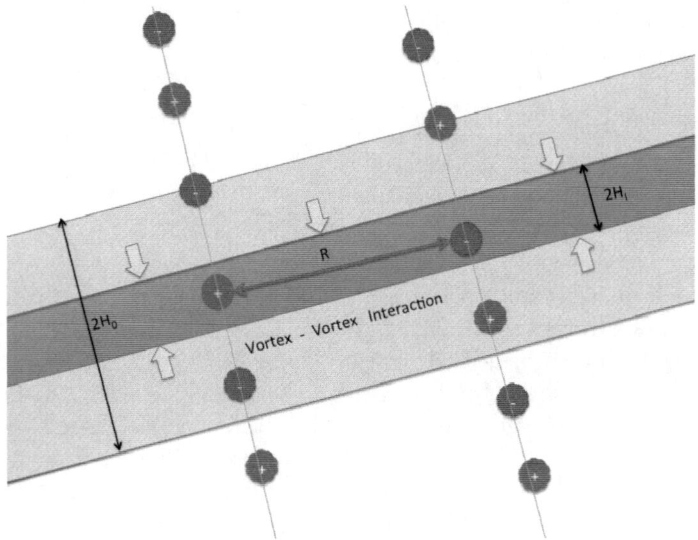

Figure 2. The logarithmic interaction of the vortex to the boundary far outweighs the vortex-vortex or vortex-anti-vortex interaction in the channel - the latter arises only on very short distances, smaller than the width of the channel. On larger distances, it weakens very fast as it is effectively the interaction between two charges, each distributed over a line of infinite length. With increasing temperature, the effective width of the channel decreases to $2H_l$ where l is the scaling parameter (see text) and so does the range of the attraction between a vortex and an anti-vortex.

The effective susceptibility $\chi = \int_{r_c}^{H-r_c} \alpha(x) dn(x)$, where $\alpha(x) = q^2(H-x)^2/2k_BT$ is the polarizability and $n(x)$ is the number density of vortices. It is straightforward to show that $dn(x) = dx \exp(-U(x)/k_BT)/(2\pi r_c^2 H)$, where $U(x)$ is the screened interaction. In terms of these, Eq. (5) reads

$$\varepsilon = 1 + \frac{q^2}{k_BT}\frac{1}{r_c^2 H}\int_{r_c}^{H-r_c} dx\,(H-x)^2 \exp(-U(x)/k_BT).$$

Introducing the dimensionless superfluid density $K = q^2/(\pi k_BT)$ and the renormalized density $K_r = K/\varepsilon(r)$, the above equation becomes an equation for K_r,

$$K_r^{-1} = K^{-1} + \frac{\pi\tilde{y}_0}{r_c^2 H}\int_{r_c}^{H-r_c} dx\,(H-x)^2 \exp\left[-2\pi K \ln\frac{2\cos kx}{kr_c}\right], \tag{3}$$

where $\tilde{y_0} = \exp(-E_c/k_BT)$. In deriving this equation we have implicitly assumed a rather low density of vortices, evident, for example, in the fact that we have used the unrenormalized charge q instead of $q_r = q/\varepsilon(r)$ to determine the polarizability. We have also neglected the correction term in the interaction energy and used the bare energy $U_0(x)$. These approximations do not change the results and are necessary to prevent the equations from becoming intractable. Strictly, one needs to replace K in the exponential by the renormalized density K_r to make Eq. (3) self-consistent. However, at low temperatures, the integral is small, and Eq. (3) is the first two terms in the expansion of K_r^{-1}.

For temperatures in the neighbourhood of the cross-over where the superfluid density tends to zero, the perturbative expansion is not valid and here we follow José *et al.* [20] and use their

IOP Conf. Series: Journal of Physics: Conf. Series **1041** (2018) 012003 doi:10.1088/1742-6596/1041/1/012003

vortex-core rescaling technique. Eventually, one arrives at the differential renormalization group equations (see, for example, [16])

$$\begin{aligned} \frac{dK_l}{dl} &= K_l - \pi\frac{(H-r_c)^2}{r_c H}K_l^2 y_l \\ \frac{dy_l}{dl} &= (1-2\pi K_l)y_l \end{aligned} \tag{4}$$

for the effective couplings K_l and y_l, together with $dH_l/dl = -H_l$. Here $dl = \ln b$ and K_l, y_l and $2H_l$ are the dimensionless density, the fugacity and the width of the channel at scale l. Note that in equations (4) H and r_c have their original bare values. For convenience, in numerical simulations presented we measure H in units of r_c: then the equations becomes dimensionless. For illustration (Fig. 3) we present the solutions to these equations for a superconducting strip of width $H = 20r_c$. We find here that there are two different flows: one flow is going to decreasing vortex fugacity and increasing superconducting density. Therewith this flow recovers the superconducting state. The other flow is going on to increasing vortex fugacity and decreasing superconducting density and therewith recovers the normal state.

The critical (or fixed) point of the scaling equations is given by $dK_k/dl = 0$ and $dy_l/dl = 0$. These two equations have the nontrivial solution

$$K^* = \frac{1}{2\pi} \text{ and } y^* = \frac{2r_c H}{(H-r_c)^2}, \tag{5}$$

describing the critical point which is a saddle point (see the renormalisation group flows presented in Fig. 3). Thus, this critical point separates the superconducting and normal states in the narrow channel. Interesting enough is the fact that the critical superconducting density does not depend on the channel width, whereas the critical value for the vortex fugacity does depend on it. For example, for a wide strip the value for critical fugacity decreases as $y^* \sim 2r_c/H$, while it increases rapidly when the channel width approaches the coherence length.

3. Vanishing of the superconducting state and the scaling behaviour in the vicinity of the critical point: results

Now we are in the position to investigate how the superconducting state vanishes. For this purpose we have to analyse the scaling in the vicinity of the critical point and connect this scaling with the thermodynamic behaviour of the system. To investigate the behaviour of the scaling, we write the scaling equations (4) in terms of scaled deviations from the fixed points by expanding K_l and y_l around (K^*, y^*) as $K_l = K^*(1+k')$ and $y_l = y^*(1+y')$. To first order in k' and y', the scaling equations then read (we now drop the primes)

$$\begin{pmatrix} \dot{k} \\ \dot{y} \end{pmatrix} = \begin{pmatrix} -1 & -1 \\ -1 & 0 \end{pmatrix} \begin{pmatrix} k \\ y \end{pmatrix}, \tag{6}$$

where $\dot{k} = dk/dl$ and similarly for $\dot{y}$. Expanding k and y in eigenstates $A_\pm(l) = A_\pm e^{\lambda_\pm l}$ of the fixed–point stability matrix above, the eigenvalues are $\lambda_\pm = \pm(\sqrt{5}\mp 1)/2$. These define the relevant and irrelevant axes in the $K_l - y_l$ plane. We assume, following existing procedure, that the relevant scaling field A_+ is the temperature axis, $A_+ \approx A|\epsilon|$, where $\epsilon = (1 - T/T_c)$ is the deviation of the temperature T from the bulk transition temperature T_c, and A is a constant.

The rescaling law for the free energy F per unit area implies

$$\begin{aligned} Z(K_0, y_0, H) &= e^{-(F_l-F_0)L^2} Z(K_l, y_l, H_l) \\ &= e^{-(F_l-F_0)L^2} Z\left(A|\epsilon|e^{\lambda_+ l}, A_- e^{\lambda_- l}, He^{-l}\right), \end{aligned} \tag{7}$$

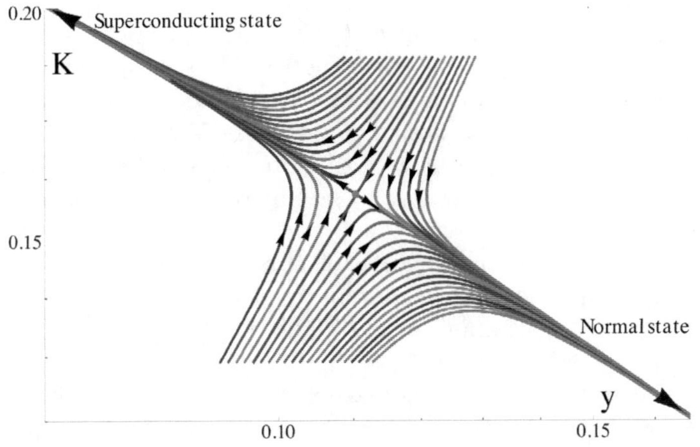

Figure 3. The scaling flows of the superconducting density K and the vortex fugacity y. All flows are split into two classes associated with the superconducting and the normal states. They are separated by the fixed point represented by the red bold dot. The flows into superconducting state are characterised by vanishing fugacity y and growing superconducting density K, while flows into the normal state are characterised by vanishing superconducting density and growing vortex fugacity.

where Z is the partition function and K_0 and y_0 are the initial values of K_l and y_l at scale size r_c. The scaling stops when $H_l = He^{-l}$ reaches a critical width $H_c = \xi = r_c e^l$. This happens when

$$l = l_c = \frac{1}{2} \ln \frac{H}{r_c}. \tag{8}$$

Setting $l = l_c$ in Eq. (7), the partition function is well defined only if

$$\xi(= r_c e^{l_c}) = r_c |\epsilon|^{-1/\lambda_+}. \tag{9}$$

Combining Eqs. (8) and (9) gives $|\epsilon|^2 = (r_c/H)^{\lambda_+}$, whence we obtain a *new* critical temperature

$$T^R = T_c \left[1 - \left(\frac{r_c}{H} \right)^{\lambda_+/2} \right], \quad \left(\frac{\lambda_+}{2} \approx 0.31 \right). \tag{10}$$

This result implies that the superfluidity breaks down at a temperature which is lower than the bulk critical temperature T_c. The mechanism of vortex nucleation we have described leads to a depression of T_c by an amount which varies inversely with the width of the superfluid channel. Previously we had done a very rough calculation on the same system (a quasi-1D superfluid channel) where we only considered two image anti-vortices (one beyond the boundary at $x = -H$ and the other beyond that at $x = +H$) and we obtained

$$T^R = T_c \left[1 - \left(\frac{r_c}{H} \right)^{1/2} \right]$$

for the temperature at which the superfluidity is destroyed [21]. Two image anti-vortices are blatantly not enough to satisfy the boundary condition of no normal flow at the boundaries of the system.

4. Conclusions/Discussion

We have described a mechanism of vortex nucleation in superfluid systems of size comparable to, or larger than, the characteristic size (coherence length) of the quasiparticles of the systems. A vortex created in a small superfluid system is strongly attracted to the boundary of the system and hence cannot nucleate. The attraction is due to the Coulomb attraction of the "vortex charge" to the "vortex charge" of the image anti-vortex. However, a fluctuating creation of several of these topological defects in the bulk of the superfluid, in the space between the vortex and the boundary, screens the V-IA attraction by renormalizing the superfluid density. This further improves the condition for the creation of more of such fluctuations. The screening effect of the plasma of vortex fluctuations continues until vortices nucleate freely. This leads to a cross-over at which the order parameter associated with the superfluidity of the system is destroyed.

We have used a real–space renormalization group method to derive the scaling laws in the vicinity of this cross-over, and shown that the cross-over occurs at a temperature which is lower than the bulk transition temperature T_c. The amount by which T_c is lowered is equal to $\Delta T_c \sim T_c(r_c/H)^a$ (with $a \approx 1/3$), showing that the depression in the transition temperature increases with decreasing channel width H.

Note that we have considered channels with widths larger than the coherence length, *i.e.*, $H > \xi$. In this regime the critical temperature decreases as the width decreases. It reaches a minimum when the width is of the order of the coherence length. When the width becomes smaller than the coherence length, the vortices cannot nucleate any more in the channel; therefore with the next decrease of the width the critical temperature may increase. However, at widths much smaller than the coherence length there may arise another phenomenon – quantum phase slip, corresponding to the vortex tunnelling from one edge of the channel to the other – an activity which is now seeing explosive interest [7, 22, 23, 24, 25, 26, 27].

The phenomenon described above is very general. Obviously it may arise in any small or porous system, like superfluid droplets, superfluid in pores of aerogels, superconducting quantum dots and so on. It may arise not only in superfluid or superconducting systems, but also in other condensed states, such as the magnetic state. In any case, topological vortex-like defects originating from the surface/boundary (like vortices in superfluids and superconductors) have the potential to destroy a condensed state provided that the system has a small size.

In this paper, we have presented a new effect that suppresses superconductivity/superfluidity in confined systems and driven by the boundary of the systems. More precisely, this suppression, although similar to the BKT transition, is driven by the creation of a plasma between the vortex and the boundary which screen the Coulomb attraction of the vortex to the boundary. In this respect, the driving mechanism is analogous to the mechanism proposed by Kusmartsev [28] for the nucleation of vortices in a flow of rotating superfluid ^{4}He, where, instead of imaginary vortices and anti-vortices, imaginary half-vortex loops play a crucial role in the destruction of the superfluid state.

Here we have discussed the thermodynamics aspect of the new low temperature physics in narrow channels where each vortex is attached to an infinitely many image vortices and image anti vortices created due to the two channel boundaries. We expect that the phenomenon we have described is more general and also exists in both equilibrium and non-equilibrium physics and dynamics of vortices. It will have a strong influence on the phase slip phenomena as well as on the electrical current propagation through superconducting wires and in superfluid channels. We intend to investigate these new phenomena in the future studies.

References

[1] Berdiyorov G R, Milosevic M V and Peeters F M 2009 *Phys. Rev. B* **79** 184506; *ibid.* **80** 214509
[2] Berdiyorov G R, Elmurodov A K, Peeters F M and Vodolazov D Y 2009 *Phys. Rev. B* **79** 174506

[3] Berdiyorov G R, Milosevic M V and Peeters F M 2012 *Appl. Phys. Lett.* **100** 262603
[4] Berdiyorov G R, Chao X H, Peeters F M, Wang H B, Moshchalkov V V and Zhu B Y 2012 *Phys. Rev. B* **86** 224504
[5] Clem J R, Mawatari Y, Berdiyorov G R and Peeters F M 2012 *Phys. Rev. B* **85** 144511
[6] Baranov V V, Balanov A G and Kabanov V V 2011 *Phys. Rev. B* **84** 094527
[7] Arutyunov K Y, Hongisto T H, Lehtinen J S, Leino L I and Vasiliev A l 2012 *Scientific Reports* **2** Article number: 293(2012)
[8] Zgirski M, Riikonen K-P, Touboltsev V and Arutyunov K 2005 *Nano Lett.* **5** 1029
[9] Fomin V M, Rezaev R O and Schmidt O G 2012 *Nano Lett.* **12** 1282
[10] Onsager L 1971 *Nuovo Cimento* Suppl. **6** 249
[11] Feynman R P 1955 *Progr. Low Temp. Phys.* **1** 17
[12] Vinen W F 1961 *Proc. Roy. Soc. (London)* a**260** 218
[13] Berezinskǐ V 1971 *Sov. Phys. JEPT* **32** 493
[14] Kosterlitz J M and Thouless J D 1973 *J. Phys. C* **6** 1181
[15] Kosterlitz J M and Thouless J D 1978 *Progr. Low Temp. Phys.* **VIIb** 373
[16] Sobnack M B and Kusmartsev F V 2001 *Phys. Rev. Lett.* **86** 716
[17] Williams G A 1987 *Phys. Rev. Lett.* **59** 1926
[18] Shenoy S R 1989 *Phys. Rev. B* **40** 5056
[19] P.K. Newton P K 2001 *The N-Vortex Problem: Analytical Techniques* (New York: Springer)
[20] José J V, Kadanoff L P, Kirkpatrick S and Nelson D R 1977 *Phys. Rev. B* **16** 1217
[21] Kusmartsev F V, Sobnack M B and Geim A K 2001 *Condensed Matter Theories, Vol. 16* (New York: Nova Science)
[22] Croitoru M D, Vagov A, Shanenko A A and Axt V M 2012 *Supercond. Sci. Technol.* **25** 124001
[23] Lehtinen J, Zakharov K and Arutyunov K 2012 *Phys. Rev. Lett.* **109** 187001
[24] Trezza M, Prischepa S L, Cirillo C and Attanasio C 2012 *Physica C* **479** 167
[25] Croitoru M D, Shanenko A A, Kaun C C and Peeters F M 2012 *J. Phys.: Condensed Matter* **24** 275701
[26] Kim H, Jamalic S and Rogachev A 2012 *Phys. Rev. Lett.* **109** 027002
[27] Mihajlo Vanevi M and Nazarov Y V 2012 *Phys. Rev. Lett.* **108** 187002
[28] Kusmartsev F V 1996 *Phys. Rev. Lett.* **76** 1880

IOP Conf. Series: Journal of Physics: Conf. Series **1041** (2018) 012004 doi:10.1088/1742-6596/1041/1/012004

Dimerization in the anisotropic bilinear-biquadratic Heisenberg model

Michael Weyrauch[1] and Mykhailo V. Rakov[2]

[1] Physikalisch-Technische Bundesanstalt, Bundesallee 100, D-38116 Braunschweig, Germany
[2] Kyiv National Taras Shevchenko University, 64/13 Volodymyrska Street, Kyiv 01601, Ukraine

E-mail: michael.weyrauch@ptb.de

Abstract. Two recent publications report different boundaries for the dimerized phase of the bilinear-biquadratic spin-1 Heisenberg model with quadratic Zeeman anisotropy (see Refs. [9] and [10]). We numerically address these discrepancies for the purely *biquadratic* model with quadratic Zeeman term using methods based on matrix product states (MPS). We extrapolate the results beyond the purely biquadratic model in order to determine the phase boundaries of the dimerized phase and propose a modified phase diagram. Our MPS implementation incorporates SU(2) symmetry explicitly and we discuss its implementation into the MPS representation in the presence of the Zeeman anisotropy.

1. Introduction

The anisotropic Heisenberg model is an immensely useful model for the study of many body (spin) systems. For a one-dimensional lattice with N sites it is given by

$$h_{i,i+1} = \cos\theta\, \vec{S}_i \otimes \vec{S}_{i+1} + \sin\theta\, (\vec{S}_i \otimes \vec{S}_{i+1})^2 + D\,(S_i^z)^2 \tag{1}$$

and includes an exchange interaction between sites i and $i+1$ controlled by the parameter θ as well as Zeeman anisotropy with strength D. The S_i^ν are spin-s SU(2) matrix representations with $\nu = x, y, z$, and $i = 1, \dots N$ and $N+1 \to 1$ for periodic boundry conditions.

This model has been studied extensively using field theoretical and numerical methods as a function of the parameters θ and D, and it can be experimentally realized in various crystalline materials as well as using trapped ultracold gases. In particular, the model was instrumental to establish the importance of topological terms in low energy field theories: For the antiferromagnetic bilinear model ($\theta = 0$) Haldane [1, 2] found a surprising difference between integer and half-interger spin chains, which was later related to a topological ϑ-term in the low energy field theory. More recently, a topological analysis [3, 4, 5] was undertaken for the dimerized region ($-3\pi/4 < \theta < -\pi/4$), which for a long time created quite some confusion with regard to a possible nematic phase as conjectured by Chubukov [6] for θ close to the transition point at $\theta = -3\pi/4$.

In order to establish and clarify such rather non-trivial features of the model, numerical investigations are absolutely essential: Most prominently, the development of the DMRG method [7] in the early nineties proved very useful in order to support the results obtained using field theoretical methods, e.g. the celebrated Haldane gap in spin-1 chains [8]. In this respect, it is rather interesting to note that the extension of the dimerized phase in the (θ, D)-plane is still rather controversial. In fact, two recent numerical investigations [9, 10] find very different extensions of the dimerized phase as a function of θ and the Zeeman strength D. We addressed these discrepancies in a recent work [11] using a matrix product state (MPS) algorithm for periodic boundary conditions with consideration of SU(2) symmetry [12, 13, 14]. In particular, we made extensive calculations at $\theta = -\pi/2$ and developed from these results a modified phase diagram for the dimerized region. In the present note we will extend these investigations and discuss the resulting phase diagram and, furthermore, comment on the numerical method and its numerical complexity.

At $\theta = -\frac{\pi}{2}$ and $D = 0$ only the biquadratic term remains in (1) which is SU(3) symmetric [15, 16], i.e. it may be rewritten as a bilinear model in terms of the three-dimensional Gell-Mann SU(3) 'quark' (λ) and 'antiquark' ($\bar{\lambda}$) triplet representations,

$$h_{i,i+1} = -8\vec{\lambda}_i \otimes \vec{\bar{\lambda}}_{i+1} - \frac{4}{3}\mathbb{1}. \tag{2}$$

The quadratic Zeeman term, which in terms of Gell-Mann matrices is given by $2D(1/3\,\mathbb{1} + \lambda_3 - \lambda_8)$, reduces the symmetry from SU(3) to SU(2), i.e. the Gell-Mann triplet splits into an SU(2) spin-$\frac{1}{2}$ duplet and one singlet at each site. We shall call this SU(2) subgroup v-spin. This SU(2) symmetry holds only at $\theta = -\pi/2, \pi/2, -3\pi/4$, and $\pi/4$, and is different from the $D = 0$ SU(2) symmetry of the Hamiltonian (1), which we shall call s-spin. The latter reduces to U(1) at all θ due to the Zeeman term. Since a continuous symmetry cannot be broken in one dimension [17, 18], we developed a matrix-product algorithm which incorporates v-spin symmetry explicitly in the ansatz for the MPS similar to our treatment of SU(2) symmetric MPS presented in Ref. [14]. Details will be discussed in section 2.

By now it is accepted [5] that the *isotropic* ($D = 0$) bilinear-biquadratic spin-1 system is dimerized for all θ between the two critical points $\theta = -\frac{3\pi}{4}$ and $\theta = -\frac{\pi}{4}$. Using the results at $\theta = -\pi/2$ as a guide, we develop and discuss in section 3 a modified phase diagram for the *anisotropic* model in the dimerized region.

2. SU(2) symmetric algorithm for v-spin

As was mentioned above, at $\theta = -\pi/2$ the *anisotropic* bilinear-biqudratic Heisenberg model is SU(2) symmetric. This is not immediately obvious from the standard spin representation given in Eq. (1), but can be understood from the SU(3) representation given in Eq. (2), which contains v-spin as a subgroup. The Zeeman term reduces the SU(3) symmetry to this SU(2) subgroup and, as a consequence, the SU(3) multiplets split into SU(2) multiplets, e.g. an SU(3) triplet splits into a SU(2) doublet and a singlet, and an SU(3) octet splits into 2 singlets, two doublets and one triplet. Therefore, the obtained states may be labeled by SU(2) v-spin quantum numbers. The U(1) subgroups of v-spin and s-spin are related by $S_z = 2v_z$. This splitting of the SU(3) multiplets can be seen in calculated spectra as exemplified in Fig. 1.

In another recent paper [14] we described in detail an algorithm for SU(2) symmetric matrix product states (MPS) with periodic boundary conditions. It is constructed from tensors separated into structural and degeneracy parts, C and $\mathcal{M}$, respectively,

$$M^{sm_s}_{(jtm),(j't'm')} = \mathcal{M}^s_{(jt),(j't')} C^{j,s,j'}_{m,m_s,m'}. \tag{3}$$

In order to form the MPS, N such tensors are contracted over the virtual spin indices (jtm) and $(j't'm')$. The (variational) parameters of the MPS are the matrix elements of the degeneracy tensors $\mathcal{M}$, while the structural part is given by a Clebsch-Gordan coefficient as detailed in Ref. [14]. Analogous expressions are defined for matrix product operators (MPO). The goal of the MPS algorithm we developed is the explicit elimination of all structural tensors. This is

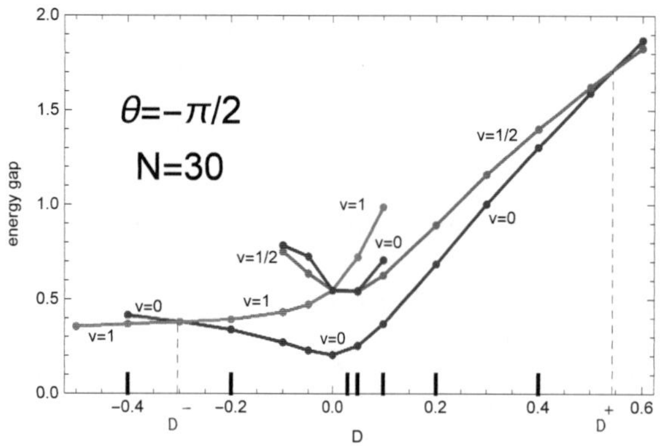

v	D	T
0	$0.2 \lesssim D \lesssim 0.6$	$\{7,7,7,0,0\}$
0	$-0.3 \lesssim D \lesssim 0.2$	$\{6,6,6,6,0\}$
0	$-0.5 \lesssim D \lesssim -0.3$	$\{5,5,5,5,5\}$
1/2	$0 \lesssim D \lesssim 0.6$	$\{7,7,7,7,0\}$
1	$-0.4 \lesssim D \lesssim 0$	$\{6,6,6,6,6\}$

Figure 1. (left) Low lying spectrum of the biquadratic ($\theta = -\pi/2$) Heisenberg ring with $N = 30$ spins and quadratic Zeeman term in the parameter range $-0.5 < D < 0.6$. The two lowest $v = 0$ excitations and the lowest $v = 1/2$ and $v = 1$ multiplets are shown (the ground state is a singlet shifted to $E = 0$). At $D = 0$, one observes one SU(3) octet above the two low lying SU(3) singlets, which splits into SU(2) v-spin multiplets at $D \neq 0$. There are two characteristic level crossings at $D^- \simeq -0.30$ and $D^+ \simeq 0.54$ indicated by dashed black vertical lines. The essential structure of the spectrum remains very similar for larger systems due to v-spin symmetry. (right) Degeneracy sets T used for the calculations of the states displayed on the left as described in the main text.

achieved by summing over all spin projection quantum numbers analytically. Details are given in Ref. [14].

In particular, for a spin-1 ring, the physical spin s of the tensor (3) takes the value 1, and the structural part C for each possible combination of quantum numbers j, j' forms a SU(2) spin-1 matrix representation. However, for v-spin we do not place a spin-1 representation at each site but the direct sum of a spin-1/2 and a spin-0 representation. This follows from the fact that v-spin is a subgroup of SU(3) with a Gell-Mann triplet as basic representation. This fact has a number of consequences for the MPS algorithm, and a slight modification to the algorithm described in Ref. [14] is required. These changes are rather technical and we do not describe the details here.

We just want to mention some essential points: while the general structure of the MPS given in Eq. (3) can be maintained, it is necessary to introduce additional loops over s with $s = 1/2$ and $s = 0$. Secondly, one has to slightly modify the definition of the representation for the matric product operator MPO (see Eq. (19) in Ref. [14]) again in order to take care that it now has off-diagonal elements in s and s'. Finally, one has to develop the explicit reduced representation of the MPO for the bilinear-biquadratic Heisenberg model including the Zeeman term D which would replace the explicit representations given in Appendix A of Ref. [14]. With these considerations in mind, it is not difficult to fill in all technical details and construct the reduced MPO representation and the corresponding algorithm.

Finally, a few remarks about the efficiency of the calculations: The complexity of the calculations is essentially determined by the choice of the quantum numbers (virtual indices) $(jt), (jt')$ with $j, j' \in \{0, 1/2, 1, 3/2, 2, \ldots\}$ and $t \in \mathbb{N}$. Here, t and t' are the degeneracies of j and j' we choose for our MPS representation. *A priori* these degeneracies are not restricted, and there is no upper limit for j or j', respectively. They have to be chosen as small as possible but such that the algorithm converges, and this choice is partly a matter of trial and error. A particular choice we call a *degeneracy set* $T = \{t_0, t_1, \ldots, t_{2j_{\max}}\}$, and for the calculations of the spectra shown in Fig. 1 they are listed in the table shown to the right of the figure.

Spectra for systems with more than $N = 30$ spins look rather similar as in Fig. 1, of course

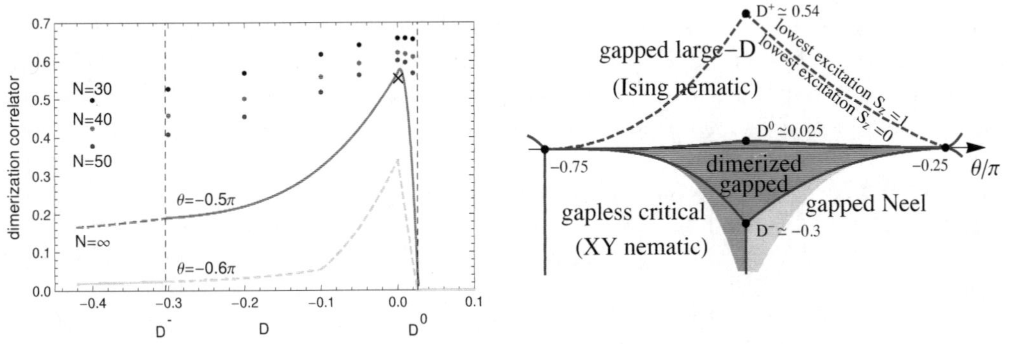

Figure 2. Dimerization (left) and schematic phase diagram (right) of the bilinear-biquadratic Heisenberg model with quadratic Zeeman anisotropy D in the range $-\frac{3\pi}{4} < \theta < -\frac{\pi}{4}$. Details are discussed in the main text.

the gaps get smaller, and with increasing system size the lowest gap(s) may close. In fact, for dimerization to happen, the lowest states must be degenerate singlets. And as we have shown in detail in our previous publication we find degenerate singlets or degenerate singlets together with a triplet for $D < .025$. For $D > .025$ the gap does not close, and the system is not dimerized.

3. Boundaries of the dimerized phase of the Heisenberg model with Zeeman anisotropy

The boundaries of the dimerized phase have been studied by Rodriguez *et al.* [9] using level spectroscopy and by De Chiara *et al.* [10] from a direct calculation of the dimerization order parameter. The results are not in agreement. In our recent paper [11] we addressed these discrepancies and pointed out the basic reason why level spectroscopic results from finite system calculations as presented in Ref. [9] are problematic: Dimerization only occurs if the lowest two singlets are degenerate in the thermodynamic limit. However, we find that the gap between the lowest two singlets does not close for all D larger than some small value close to zero. More details may be found in Ref. [11].

Here we concentrate on calculations of the dimerization order parameter $\hat{\mathcal{D}} = \frac{1}{N}\sum_i (-1)^i\, h_{i,i+1}$. Results for the dimerization correlator are presented in Fig. 2 (left). For $D = 0$ the dimerization is well-known from the literature [19, 20], $\mathcal{D}_\infty = \frac{\sqrt{5}}{2}\prod_{n=1}^{\infty}\tanh^2(n\,\mathrm{arccosh}\frac{3}{2}) \simeq 0.562$. Our results for 30, 40 and 50 sites at $D = 0$ can be fitted very well by the function [21, 22] $\mathcal{D}(N) = \mathcal{D}_\infty + c\,N^{-\alpha}\exp(-N/2\xi)$ with $\alpha = 1$. From the fit we obtain $\mathcal{D}_\infty \simeq 0.568$ in good agreement with the Bethe Ansatz, and a large correlation length $\xi \simeq 20.2$. A very similar result for the correlation length was obtained in Ref. [21] from the lowest energy gap.

The dimerization correlator is determined from the two lowest $v = 0$ eigenstates with different momenta for 30, 40, and 50 sites. The red line in Fig. 2 (left) shows the extrapolation to the thermodynamic limit. The black cross indicates the Bethe ansatz value at $D = 0$ and $N \to \infty$. The dimerization is strictly zero for $D \gtrsim 0.03$. The green line shows the result obtained by De Chiara *et al.* [10] at $\theta = -0.6\pi$ (the dashed part is extrapolated from their data). Our results confirm the prediction of De Chiara *et al.* that the dimerization persists up to large negative values of D.

Let us now transfer these results at $\theta = -\pi/2$ into the phase diagram shown in Fig. 2 (right) and combine them with insights we obtained from the spectral analysis of Ref. [11]. We find

IOP Conf. Series: Journal of Physics: Conf. Series **1041** (2018) 012004 doi:10.1088/1742-6596/1041/1/012004

that a *gapped* dimerized phase exists in the parameter range $D^- < D < D^0$ with $D^- \simeq -0.30$ and $D^0 \simeq 0.025$. Moreover, we identify a *gapless* dimerized phase which extends from D^- to large negative D. While this confirms the results of De Chiara *et al.* [10], who predicted a small dimerization even below $D < -2.0$, the existence of both a gapped and a gapless dimerized phase is reported here for the first time. The transition between these two dimerized regions occurs at D^- which was erroneously identified as the boundary of the dimerized phase in Ref. [9]. At the upper end of the dimerized region close to D^0, the dimerization sharply drops to zero and a gap opens between the two lowest singlet states marking the transition to a non-dimerized phase. We do not see a phase transition at D^+ which was identified as the upper dimerized phase boundary in Ref. [9]. These findings for $\theta = -\frac{\pi}{2}$ are graphically represented on the vertical axis of the phase diagram shown in Fig.2 (right), where the various transition points are marked by black dots.

We now qualitatively extrapolate these results for θ in the parameter interval $I = [-\frac{3\pi}{4}, -\frac{\pi}{4}]$, separately for positive D and negative D, guided by general considerations and the calculations presented in Refs. [9] and [10]: By now it is rather well established that for $D = 0$ the bilinear-biquadratic spin-1 model has a dimerized ground state in the whole parameter interval I. In particular, a nematic non-dimerized phase close to the ferromagnetic transition has been ruled out [4, 5]. At large negative or positive D the system is not dimerized. This follows from simple analytical arguments [10].

At large positive $D \gg 1$ the system is in the gapped large-D phase [10], and the transition between the dimerized phase to a non-dimerized phase happens at small positive D for all $\theta \in I$ [10]. This we confirmed in the present paper for $\theta = -\pi/2$. Furthermore, we recently made a number of calculations at various θ, which clearly show that the transition to the dimerized phase is at $D < 0.02$. In fact, one expects that the system is in an Ising nematic phase for all $D > D^0$ as indicated by the white region above the blue shaded region in Fig. 2 (right) as there are no gaps closing in the spectrum. However, it is expected from the results of Ref. [9] that the leading excitation changes from $S_z = 0$ for $D < D^+$ to $S_z = \pm 1$ at $D > D^+$ as indicated by the blue dashed lines in the phase diagram. According to Ref. [5] the dimerization is related to the density of disclinations created in the spin system. Consequently, such topological defects should be absent for $D > D^0$.

For large negative D the system is in a gapless critical (XY nematic) phase for $-\frac{3\pi}{4} < \theta < -\frac{\pi}{2}$ and in a gapped Neel phase for $-\frac{\pi}{2} < \theta < \frac{\pi}{2}$ [10]. For small and intermediate negative D, the XY nematic and the Neel phases are separated by dimerized phases as indicated by the blue, red, and green shaded regions in Fig. 2 (right). It is expected that the gap between the lowest two singlets closes in all theses colored regions making them dimerized. In addition, in the red region also the gap to the next triplet closes, i.e. one expects a 5-fold degenerate ground state and vanishing staggered magnetization. This corresponds to our findings at $\theta = -\frac{\pi}{2}$. In the dimerized green shaded region, we expect an open gap to the triplet state and a non-zero staggered magnetization. The line between the red and green dimerized sectors separates magnetically staggered and non-staggered phases. This must be confirmed in detail by further calculations.

4. Conclusions

In this paper we elaborated on our proposal [11] for a modified phase diagram in the dimerized region of the bilinear-biquadratic spin-1 Heisenberg model and resolved discrepancies for the phase boundaries reported in previous work [9, 10]. The modified phase diagram is presented in our Fig. 2, which indicates that for positive anisotropy paramater D the system is not dimerized except at very small positive D. Our proposal is based on numerical results from matrix product state calculations, which explicitly maintain SU(2) symmetry.

Acknowledgments
We thank Alexei K. Kolezhuk for discussions. Mykhailo V. Rakov thanks Physikalisch-Technische Bundesanstalt for financial support during short visits to Braunschweig.

References

[1] Haldane F D M 1983 *Phys. Lett. A* **93** 464 ISSN 0375-9601
[2] Haldane F D M 1983 *Phys. Rev. Lett.* **50**(15) 1153–1156
[3] Ivanov B A and Kolezhuk A K 2003 *Phys. Rev. B* **68**(5) 052401
[4] Grover T and Senthil T 2007 *Phys. Rev. Lett.* **98**(24) 247202
[5] Hu S, Turner A M, Penc K and Pollmann F 2014 *Phys. Rev. Lett.* **113**(2) 027202
[6] Chubukov A V 1991 *Phys. Rev. B* **43**(4) 3337
[7] White S R 1992 *Phys. Rev. Lett.* **69**(19) 2863
[8] White S R and Huse D A 1993 *Phys. Rev. B* **48** 3844
[9] Rodríguez K, Argüelles A, Kolezhuk A K, Santos L and Vekua T 2011 *Phys. Rev. Lett.* **106**(10) 105302
[10] De Chiara G, Lewenstein M and Sanpera A 2011 *Phys. Rev. B* **84**(5) 054451
[11] Weyrauch M and Rakov M V 2017 *Phys. Rev. B* **96**(13) 134404
[12] Weyrauch M and Rakov M V 2013 *Ukr. J. Phys.* **58** 657
[13] Rakov M V, Weyrauch M and Braiorr-Orrs B 2016 *Phys. Rev. B* **93** 054417
[14] Rakov M V and Weyrauch M 2017 *J. Phys. Commun.* **1** 015007
[15] Affleck I 1986 *Nuclear Physics B* **265** 409 – 447 ISSN 0550-3213
[16] Batista C D, Ortiz G and Gubernatis J E 2002 *Phys. Rev. B* **65**(18) 180402
[17] Mermin N D and Wagner H 1966 *Phys. Rev. Lett.* **17**(22) 1133
[18] Coleman S 1973 *Comm. Math. Phys.* **31** 259 ISSN 1432-0916
[19] Xian Y 1993 *Phys. Lett. A* **183** 437 ISSN 0375-9601
[20] Baxter R J 1973 *J. Stat. Phys.* **9** 145 ISSN 0022-4715
[21] Sørensen E S and Young A P 1990 *Phys. Rev. B* **42**(1) 754
[22] Buchta K, Fáth G, Legeza O and Sólyom J 2005 *Phys. Rev. B* **72**(5) 054433

Advances in non-relativistic matter via complex Langevin approaches

Joaquín E. Drut

Department of Physics and Astronomy, University of North Carolina, Chapel Hill, North Carolina, 27599-3255, USA

E-mail: drut@email.unc.edu

Abstract. The recent progress in understanding the mathematics of complex stochastic quantization, as well as its application to quantum chromodynamics in situations that have a complex phase problem (e.g. finite quark density, real time), has opened up an intriguing possibility for non-relativistic many-body physics which has so far remained largely unexplored. In this brief contribution, I review a few specific examples of advances in the characterization of the thermodynamics of non-relativistic matter in a variety of one-dimensional cases affected by the sign problem: repulsive interactions, finite polarization, finite mass imbalance, and projection to finite systems to obtain virial coefficients.

1. Introduction

This contribution is concerned with computational approaches to the generic quantum many-body problem, as it appears in high-energy, nuclear, condensed matter, and atomic physics; and quantum chemistry. The standard non-perturbative computational methods to this problem can be roughly divided into two large sets: direct methods and stochastic methods [which we will collectively refer to below as quantum Monte Carlo methods (QMC)]. In the former category we find exact diagonalization and configuration interaction approaches. While these are very appealing due to their potential for high precision, their memory requirements (especially in 3 spatial dimensions) can quickly become prohibitive as the system size is increased. Stochastic methods, which are the focus of this work, tend to have much milder memory requirements but suffer from other challenging issues.

Indeed, by far, most interesting problems in quantum many-body physics suffer from the infamous sign problem when one attempts to compute, using stochastic methods, essentially any useful quantity, be it ground-state properties, thermodynamics, or dynamic response (see e.g. [1, 2]). The sign problem is a severe signal-to-noise issue characterized by an exponentially decreasing signal-to-noise ratio as the volume of spacetime is increased (see e.g. [3, 4]). The hallmark of the sign problem can often be detected without doing a single calculation: the probability measure in the path integral may be real but of indefinite sign, or it may be complex, in which case we speak of a 'phase problem'.

In spite of the general severity of the problem, in some cases it is possible to carry out calculations using a simple re-weighting technique, described next (see also e.g. Ref. [5]). Starting from the path integral formulation of the quantum many-body problem, where the partition function takes the form (see below for further details on a particular case)

$$\mathcal{Z} = \int \mathcal{D}\sigma \, e^{-S[\sigma]}, \tag{1}$$

where the action $S[\sigma]$ is generally complex, which complicates the identification of $P[\sigma] = e^{-S[\sigma]}$ as a probability. For example, in the presence of fermionic degrees of freedom (electrons in condensed matter, atoms in ultracold atom experiments, quarks in QCD), one has

$$P[\sigma] = \det M[\sigma], \tag{2}$$

where $M[\sigma]$ contains the dynamics of the system and is generally a complex matrix.

The expectation value of an operator $\hat{O}$ will generally assume the schematic form

$$\langle \hat{O} \rangle = \frac{\int \mathcal{D}\sigma P[\sigma] O[\sigma]}{\int \mathcal{D}\sigma P[\sigma]}, \tag{3}$$

where $P[\sigma]$ is, as mentioned above, what one would normally want to identify as a probability measure (usually a fermion determinant and other pieces of the action), but which will typically be complex. Rearranging the above expression by separating the magnitude $|P[\sigma]|$ and the phase $e^{i\phi[\sigma]}$, and multiplying and dividing by a normalization factor, one obtains

$$\langle \hat{O} \rangle = \frac{\int \mathcal{D}\sigma |P[\sigma]| e^{i\phi[\sigma]} O[\sigma]}{\int \mathcal{D}\sigma |P[\sigma]|} \frac{\int \mathcal{D}\sigma |P[\sigma]|}{\sum_\sigma |P[\sigma]| e^{i\phi[\sigma]}} = \frac{\langle\langle e^{i\phi[\sigma]} O[\sigma] \rangle\rangle}{\langle\langle e^{i\phi[\sigma]} \rangle\rangle}. \tag{4}$$

In this expression, the double angle bracket denotes an expectation value with respect to the non-negative measure $|P[\sigma]|$. One may thus use $|P[\sigma]|$ as the probability in conventional Metropolis-based methods. However, the price of such a re-weighting trick can be high: both the numerator and the denominator in the above expression will approach zero exponentially (as a function of the system size, see e.g. Ref. [3]), quickly going beyond the precision of any reasonable calculation (unless astronomical numbers of samples are used), such that obtaining their ratio becomes impractical.

In spite of remarkable advances toward taming this issue (i.e. find ways to make the phase average stay as close to 1 as possible), the search for a more generic or practical solution beyond the above re-weighting idea continues. Remarkably, some lines of research have shown that in specific cases the sign problem can be completely avoided (see e.g. [1, 2]); however, we will here focus on methods that attempt to overcome the problem by implementing some form of complex stochastic quantization, also known as complex Langevin (CL), which was brought up in Refs. [6, 7]. The idea of complex stochastic quantization is, in fact, a line of research that has flourished in recent years after multiple abandoned attempts in the area of lattice QCD [8, 9]. As it appears that the understanding of CL is now far better than ever before, it is an appealing candidate for addressing the sign problem in areas outside QCD.

In this contribution we show some of the recent attempts to use CL to overcome the sign problem in a class of non-relativistic many-body problems, keeping in mind their realization in ultracold atom experiments. We begin by describing some of the operational aspects of the method, leaving the formal aspects to the excellent extant literature (see e.g. [10, 11, 12, 13, 14]). We then review recent results for spin-1/2 fermions in one spatial dimension (1D) focusing on repulsive interactions at finite temperature, spin polarized systems, and mass-imbalanced systems. Finally, we show progress towards implementing particle projection algorithms via CL, which enables the stochastic calculation of high-order virial coefficients. Note that, in the 1D cases described here, we implement contact interactions, as befits dilute gases. In some cases the sign problem can be avoided entirely by using other algorithms [15]; however, those algorithms do not generalize to higher dimensions, which makes them less interesting for our objectives.

2. Outline of formalism and methods

For reference, it is useful to recall here the basic operations involved in the hybrid Monte Carlo (HMC) algorithm [16, 17, 18, 19]. We begin by writing the grand-canonical partition function

$$\mathcal{Z} = \mathrm{tr}\, e^{-\beta(\hat{H}-\mu\hat{N})}, \tag{5}$$

as a path integral, where $\hat{H}$ is the Hamiltonian, $\hat{N}$ is the total particle number operator, β is the inverse temperature, and μ the chemical potential. Following conventional steps (see e.g. [18]), we begin by discretizing the imaginary time direction as $\beta = N_\tau \tau$ for a given time step τ and using a Suzuki-Trotter decomposition of the transfer-matrix operator, e.g.

$$e^{-\tau\hat{H}} = e^{-\tau\hat{T}} e^{-\tau\hat{V}} + O(\tau^2), \tag{6}$$

where $\hat{T}$ is the kinetic energy operator and $\hat{V}$ is the potential energy operator containing the interaction (we assume that an external potential, if present, is included in $\hat{T}$). A Hubbard-Stratonovich (HS) transformation (of which there are many choices, see e.g. [18]) then allows us to write the interaction factor as a path integral over the auxiliary field σ (which would typically be the photon field in electrodynamics and the gluon field in QCD):

$$e^{-\tau\hat{V}} = \int \mathcal{D}\sigma e^{-\tau W[\sigma]}, \tag{7}$$

where $W[\sigma]$ is simply a one-body operator containing the external auxiliary field σ. Inserting the factorization and HS-transformed expressions into the partition function, we arrive at the the path integral form mentioned above, namely

$$\mathcal{Z} = \int \mathcal{D}\sigma\, e^{-S[\sigma]}, \tag{8}$$

where the effective action $S[\sigma]$ is typically the logarithm of a fermion determinant and potentially includes pure σ terms, as in QCD (though not necessarily so in non-relativistic physics).

When $S[\sigma]$ is real, the usual methods sample the field σ according to $P[\sigma] = e^{-S[\sigma]}$. The HMC algorithm carries out that sampling in a global fashion (i.e. modifying the whole σ field at the same time) by enlarging the phase space with the introduction of an auxiliary momentum field π conjugate to σ whose dynamics factorizes in the partition function and therefore does not alter the physics of the problem. Explicitly,

$$\mathcal{Z} = \int \mathcal{D}\sigma\mathcal{D}\pi\, e^{-\mathcal{H}[\sigma,\pi]}, \tag{9}$$

where $\mathcal{H}[\sigma,\pi] = \frac{1}{2}\sum_x \pi(x)^2 + S[\sigma]$. The classical equations of motion for $\mathcal{H}[\sigma,\pi]$, in a fictitious phase-space time t, are then used to update the fields:

$$\dot{\sigma} = \pi, \tag{10}$$

$$\dot{\pi} = -\frac{\delta S[\sigma]}{\delta\sigma}. \tag{11}$$

In the above phase-space evolution, a trajectory of length $t \sim O(1)$ is considered a full update of the field (although several such trajectories are usually required to ensure decorrelation). In between trajectories, a Metropolis step is implemented to ensure that the correct distribution is being sampled, and the momentum π is refreshed using a Gaussian distribution. Note that, in such a classical evolution, the fictitious energy given by the value of $\mathcal{H}$ is conserved (as long as the integrator is accurate enough), which ensures a very high acceptance rate in the Metropolis step. The HMC algorithm thus succeeds at implementing global updates with high acceptance rate. From this brief operational description, it can be inferred that the most costly part of the algorithm is the calculation of the force $\delta S[\sigma]/\delta\sigma$.

The stochastic quantization sister of the HMC method is what one may call real Langevin (RL) [20, 21], which enables global updates as well. In RL, however, there is no Metropolis step and the equations of motion take the form

$$\dot{\sigma} = -\frac{\delta S[\sigma]}{\delta\sigma} + \eta, \tag{12}$$

where we note that there is no auxiliary momentum field π but a t-dependent noise field η appears instead. The latter satisfies $\langle\eta(x,\tau)\rangle = 0$ and $\langle\eta(x,\tau)\eta(x',\tau')\rangle = 2\delta_{x,x'}\delta_{\tau,\tau'}$ for spacetime points (x,τ) and (x',τ'), and may be chosen to be Gaussian. The similarities between RL and HMC are clear; most notably, the calculation of the 'drift' (as is often called in the context of stochastic quantization) given by $\delta S[\sigma]/\delta\sigma$, is the most computationally intensive part of the method.

The conventional mathematical underpinnings of HMC and RL depend on $P[\sigma]$ being positive semidefinite (i.e. $S[\sigma]$ being real), a property which fails to hold in non-relativistic physics for repulsive interactions (away from the half-filling point), polarized systems, mass-imbalanced systems, and so on, and the calculation is then said to have a sign problem (or more generally a complex-phase problem). In the case of HMC, this means that the Metropolis step is simply no longer well-defined and thus the algorithm is no longer available. For RL, on the other hand, a generalization is possible into what we referred to above as CL.

Operationally, in CL one complexifies the HS field σ via

$$\sigma = \sigma_R + i\sigma_I, \tag{13}$$

where σ_R and σ_I are both real fields and defines equations of motion by

$$\delta\sigma_R = -\mathrm{Re}\left[\frac{\delta S[\sigma]}{\delta\sigma}\right]\delta t + \eta\sqrt{\delta t}, \tag{14}$$

$$\delta\sigma_I = -\mathrm{Im}\left[\frac{\delta S[\sigma]}{\delta\sigma}\right]\delta t, \tag{15}$$

where now $S[\sigma]$ is to be understood as a complex function of the complex variable σ. Clearly, when the action is real, the imaginary part of the force vanishes and CL reduces to RL. We also note that we chose real noise, but complex noise is also a possibility (see e.g. Ref. [10]).

3. Results

As anticipated, this section reviews some of the recent applications of the complex Langevin method to non-relativistic systems. Four different situations are briefly described (all corresponding to fermions in 1D with a zero-range interaction): repulsive interactions, finite polarization, finite mass imbalance, and particle number projection. It should be pointed out that, while 1D systems can be tackled with the Bethe ansatz (see e.g. Ref. [22]), although the latter runs into difficulties at finite temperature and in the grand-canonical ensemble, and at finite mass imbalance. All of the results presented here correspond to nonrelativistic fermions with an interaction of the form

$$\hat{V} = \int dx\; g\hat{n}_\uparrow(x)\hat{n}_\downarrow(x), \tag{16}$$

where g is the coupling and $\hat{n}_s$ is the density operator for particles of spin s.

3.1. Spin-1/2 fermions in 1D: repulsive interactions at finite-temperature

Our first results with CL focused on the density equation of state n of unpolarized fermions with repulsive interactions at finite temperature, which were first published in Ref. [23]. This equation of state is of interest because it is relatively easy to measure in experiments (see e.g. [24, 25, 26]) and from it one may obtain the pressure by integration and the compressibility by differentiation. In short, it is the simplest observable characterizing the thermodynamics of these quantum systems.

The auxiliary field method gives a sign problem for this system regardless of the dimension, but we focused on 1D as a starting point and compared our CL results with lattice perturbation theory up to next-to-leading (NLO), next-to-next-to-leading (N2LO), and next-to-next-to-next-to-leading order (N3LO), in Fig. 1 (left panel) for several values of the coupling λ. In the region of large fugacity, i.e. $\beta\mu \gg 0$ (where β is the inverse temperature and μ the chemical potential, such that $z = \exp(\beta\mu)$ is the fugacity), we expect perturbative results to be reliable, and we find that CL agrees with them. As the coupling is increased, perturbation theory begins to break down, but CL continues to converge, giving a prediction for the grand canonical equation of state for the density.

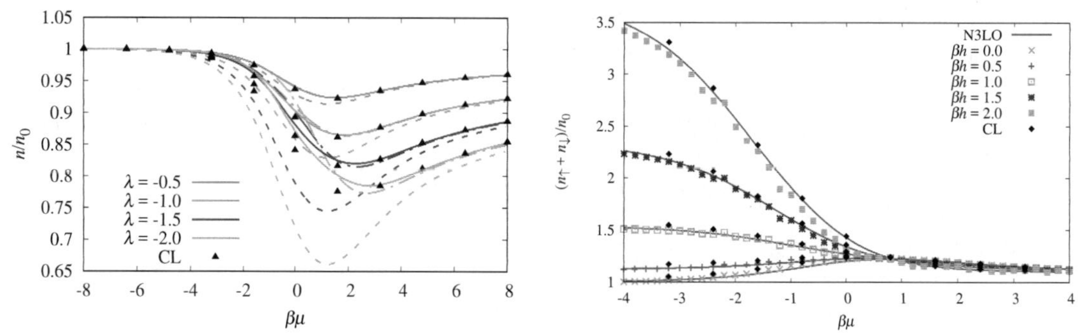

Figure 1. Left panel: Density equation of state n of 1D fermions with repulsive interactions, in units of the noninteracting counterpart n_0. Black diamonds show the CL results. Dashed, dash-dotted, and solid lines show results of NLO, N2LO, and N3LO lattice perturbation theory, respectively [23]. Right panel: Density equation of state of 1D fermions with attractive interactions at finite chemical potential asymmetry βh, which leads to different densities $n_\uparrow$ and $n_\downarrow$ for each spin species. Black diamonds show the CL results. Solid lines show results of N3LO lattice perturbation theory. Colored data points show the results of imaginary-polarization calculations of Ref. [27].

3.2. Spin-1/2 fermions in 1D: finite polarization at finite-temperature

Following up on the success of our study of repulsive interactions, we move on to polarized systems, which are especially interesting due to the possibility of realizing exotic superfluids such as the Fulde-Ferrell-Larkin-Ovchinnikov phase. As in the previous case, we compared with perturbative results, but also with the imaginary-polarization approach of Ref. [27] (which used an idea first put forward in Ref. [28]), in Fig. 1 (right panel). This case is more remarkable than the previous one because we obtain results that are not only in agreement with N3LO perturbation theory but also with the entirely different non-perturbative approach of imaginary polarization.

IOP Conf. Series: Journal of Physics: Conf. Series **1041** (2018) 012005 doi:10.1088/1742-6596/1041/1/012005

3.3. Spin-1/2 fermions in 1D: ground-state of mass-imbalanced systems

The realization of ultracold atomic gases with different atomic species yields another area of research where the sign problem prevents useful calculations. However, by implementing imaginary mass-imbalance and CL approaches [29, 30, 31], we have shown that this field is entirely open for attractive interactions and at least partially open for the repulsive case. Indeed, a signal-to-noise issue reappears due to vanishing weight $P[\sigma]$ in the complex σ plane (i.e. the appearance of singularities in $S[\sigma]$) at large repulsive interactions. This issue is currently open and under investigation. Progress in this direction is shown in the left panel of Fig. 2, where we show the ground-state energy (in units of its non-interacting counterpart) of 10 fermions, as a function of the coupling $\gamma = g/n$. The plot shows results for several mass imbalances $\bar{m} = (m_\uparrow - m_\downarrow)/(m_\uparrow + m_\downarrow)$.

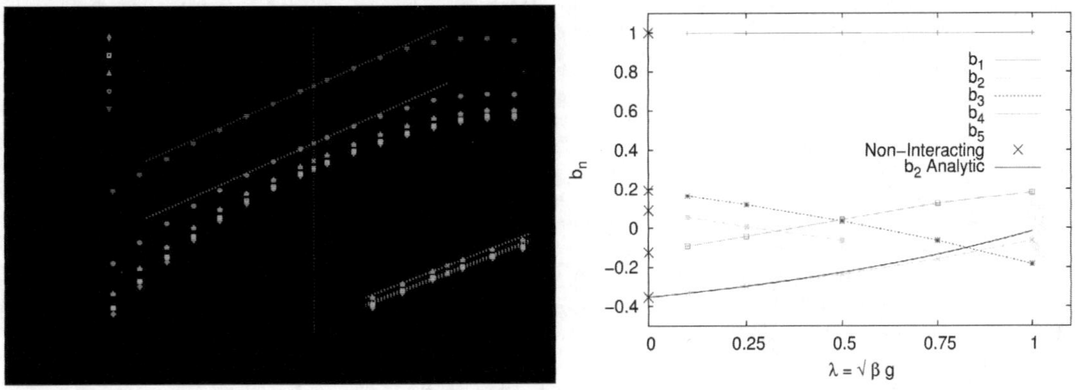

Figure 2. Left panel: Ground-state energy E of 10 mass-imbalanced fermions in 1D, as a function of the coupling $\gamma = g/n$ [31], where n is the density, in units of the noninteracting counterpart E_{FG}. Dashed lines show a first-order perturbative result. Right panel: First few virial coefficients b_n of the attractive 1D Fermi gas, obtained by Fourier-based particle projection.

3.4. Particle-projection approach to clusters and high-order virial coefficients

The idea of implementing a Fourier-transform particle-number projection of a grand-canonical calculation to obtain information about finite systems is not new; it is quite common in the context of nuclear physics (see e.g. [32]). However, to our knowledge it has not been applied by way of the complex Langevin algorithm, nor has it been applied to the stochastic extraction of virial coefficients from grand canonical calculations. Here, we report briefly on progress toward that goal, focusing on 1D systems as test cases, as done in previous subsections. In Fig. 2 we show our first results, which indicate that it may be possible to extract up to fifth order coefficients. In that figure we map out the coupling-constant dependence of those virial coefficients and compare with the analytic result of Ref. [33] for b_2. The remarkable agreement in the case of b_2 (up to systematic effects as λ increases), along with the relative smoothness of the results for the other coefficients, indicate that the method is robust, although certainly not as precise as other approaches based on the exact solution of the n-body spectrum. Note that the approach also correctly reproduces $b_1 = 1$ for all couplings. It remains an open question whether the method remains viable in higher dimensions, but a priori nothing indicates otherwise.

4. Conclusions

In this contribution we have reviewed the recent exploration of the application of complex stochastic quantization to the non-relativistic quantum many-body problem. Remarkably, a wide range of applicability was found and validated in several ways, e.g. with third-order perturbation theory or imaginary-asymmetry methods. Specifically, we were able to obtain equations of state of polarized and mass-imbalanced matter with attractive and repulsive interactions in 1D. We have also shown that practical calculations of high-order virial coefficients is possible using the idea or particle-number projection along with CL.

Acknowledgments

I would like to thank J. Braun, A. C. Loheac, L. Rammelmüller, and C. R. Shill for discussions and data. This material is based upon work supported by the National Science Foundation under Grants No. PHY1452635 (Computational Physics Program). Numerical calculations have partially been performed at the LOEWE-CSC Frankfurt.

References

[1] C. Gattringer and K. Langfeld, *Approaches to the sign problem in lattice field theory*, Int. J. Mod. Phys. A **31**, 1643007 (2016).
[2] L. Bongiovanni, *Numerical methods for the sign problem in lattice field theory*, arXiv:1603.06458.
[3] *Phase of the Fermion Determinant at Nonzero Chemical Potential*, K. Splittorff and J. Verbaarschot, Phys. Rev. Lett. **98**, 031601 (2007).
[4] *The Approach to the Thermodynamic Limit in Lattice QCD at $\mu \neq 0$*, K. Splittorff and J. Verbaarschot, Phys. Rev. D **77**, 014514 (2008).
[5] E. P. Munger and M. A. Novotny, *Reweighting in Monte Carlo and Monte Carlo RenormalizationGroup Studies*, Phys. Rev. B **43**, 5773 (1991).
[6] G. Parisi, *On complex probabilities*, Phys. Lett. B **131**, 393 (1983).
[7] J.R. Klauder, *Coherent-state Langevin equations for canonical quantum systems with applications to the quantized Hall effect*, Phys. Rev. A **29**, 2036 (1984).
[8] G. Aarts, *Introductory lectures on lattice QCD at nonzero baryon number*, J. Phys. Conf. Ser. **706**, 022004 (2016).
[9] G. Aarts and I.-O. Stamatescu, *Stochastic quantization at finite chemical potential*, J. High Energy Phys. 09 (2008) 018.
[10] G. Aarts, E. Seiler, and I.-O. Stamatescu, *The complex Langevin method: When can it be trusted?*, Phys. Rev. D **81**, 054508 (2010).
[11] G. Aarts, F. A. James, E. Seiler, and I.-O. Stamatescu, *Complex Langevin: Etiology and diagnostics of its main problem*, Eur. Phys. J. C **71**, 1756 (2011).
[12] G. Aarts, E. Seiler, D. Sexty and I. O. Stamatescu, *Complex Langevin dynamics and zeroes of the fermion determinant*, arXiv:1701.02322
[13] K. Nagata, J. Nishimura, and S. Shimasaki, *Argument for justification of the complex Langevin method and the condition for correct convergence*, Phys. Rev. D **94**, 114515 (2016).
[14] J. Nishimura, and S. Shimasaki, *New insights into the problem with a singular drift term in the complex Langevin method*, Phys. Rev. D **92**, 011501 (2015).
[15] S. Zhang, private communication.
[16] S. Duane, A. D. Kennedy, B. J. Pendleton, D. Roweth, *Hybrid Monte Carlo*, Phys. Lett. B **195**, 216 (1987).
[17] S. A. Gottlieb, W. Liu, D. Toussaint, R. L. Renken, *Hybrid-molecular-dynamics algorithms for the numerical simulation of quantum chromodynamics*, Phys. Rev. D **35**, 2531 (1987).
[18] J. E. Drut and A. N. Nicholson, *Lattice methods for strongly interacting many-body systems*, J. Phys. G: Nucl. Part. Phys. **40**, 043101 (2013).
[19] D. Lee, *Lattice simulations for few- and many-body systems*, Prog. Part. Nucl. Phys. **63**, 117 (2009).
[20] P. H. Damgaard and H. Hüffel, *Stochastic quantization*, Phys. Rep. **152**, 227 (1987).
[21] G. G. Batrouni, G. R. Katz, A. S. Kronfeld, G. P. Lepage, B. Svetitsky and K. G. Wilson, *Langevin simulations of lattice field theories*, Phys. Rev. D **32**, 2736 (1985).
[22] X-W. Guan, M. T. Batchelor, and C. Lee, *Fermi gases in one dimension: From Bethe ansatz to experiments*, Rev. Mod. Phys. **85**, 1633 (2013).
[23] A. C. Loheac and J. E. Drut, *Third-order perturbative lattice and complex Langevin analyses of the finite-*

IOP Conf. Series: Journal of Physics: Conf. Series **1041** (2018) 012005 doi:10.1088/1742-6596/1041/1/012005

temperature equation of state of nonrelativistic fermions in one dimension, Phys. Rev. D **95**, 094502 (2017).

[24] K. Van Houcke, F. Werner, E. Kozik, N. Prokofev, B. Svistunov, M. J. H. Ku, A. T. Sommer, L. W. Cheuk, A. Schirotzek, M. W. Zwierlein, *Feynman diagrams versus Fermi-gas Feynman emulator*, Nature Phys. **8**, 366 (2012).

[25] I. Boettcher, L. Bayha, D. Kedar, P.A. Murthy, M. Neidig, M.G. Ries, A.N. Wenz, G. Zürn, S. Jochim, and T. Enss, *Equation of State of Ultracold Fermions in the 2D BEC-BCS Crossover Region*, Phys. Rev. Lett. **116**, 045303 (2016).

[26] K. Fenech, P. Dyke, T. Peppler, M.G. Lingham, S. Hoinka, H. Hu, and C.J. Vale, *Thermodynamics of an attractive 2D Fermi gas*, Phys. Rev. Lett. **116**, 045302 (2016).

[27] A. C. Loheac, J. Braun, J. E. Drut, D. Roscher, *Thermal equation of state of polarized fermions in one dimension via complex chemical potentials*, Phys. Rev. A **92**, 063609 (2015).

[28] J. Braun, J.-W. Chen, J. Deng, J. E. Drut, B. Friman, C.-T. Ma, Y.-D. Tsai, *Imaginary Polarization as a Way to Surmount the Sign Problem in Ab Initio Calculations of Spin-Imbalanced Fermi Gases*, Phys. Rev. Lett. **110**, 130404 (2013).

[29] D. Roscher, J. Braun, and J. E. Drut, *Inhomogeneous phases in one-dimensional mass- and spin-imbalanced Fermi gases*, Phys. Rev. A **89**, 063609 (2014).

[30] J. Braun, J. E. Drut, D. Roscher, *Zero-temperature equation of state of mass-imbalanced resonant Fermi gases*, Phys. Rev. Lett. **114**, 050404 (2015).

[31] L. Rammelmüller, W. J. Porter, J. E. Drut, J. Braun, *Surmounting the sign problem in non-relativistic calculations: a case study with mass-imbalanced fermions*, Phys. Rev. D **96**, 094506 (2017).

[32] M. Bender, P.-H. Heenen, and P.-G. Reinhard, *Self-consistent mean-field models for nuclear structure*, Rev. Mod. Phys. **75**, 121 (2003).

[33] M. D. Hoffman, P. D. Javernick, A. C. Loheac, W. J. Porter, E. R. Anderson, and J. E. Drut, *Universality in one-dimensional fermions at finite temperature: Density, compressibility, and contact*, Phys. Rev. A **91**, 033618 (2015).

A complex Langevin approach to ultracold fermions

Lukas Rammelmüller[1], **Joaquín E. Drut**[2], **Jens Braun**[1,3]

[1] Institut für Kernphysik (Theoriezentrum), TU Darmstadt, D-64289 Darmstadt, Germany.
[2] Department of Physics and Astronomy, University of North Carolina, Chapel Hill, NC, 27599, USA.,
[3] ExtreMe Matter Institute EMMI, GSI, Planckstrasse 1, D-64291 Darmstadt, Germany.

E-mail: `lrammelmueller@theorie.ikp.physik.tu-darmstadt.de`

Abstract. The theoretical treatment of Fermi systems consisting of particles with unequal masses is challenging. Even in one spatial dimension analytic solutions are limited to special configurations and numerical progress with Monte Carlo simulations is hindered by the sign-problem. To circumvent this issue, we exploit the Complex Langevin approach and study one-dimensional mass-imbalanced two-component Fermi gases with attractive and repulsive interactions. We find perfect agreement with results obtained by other methods in a range of parameter space. Promisingly, our approach is not limited to the specific model presented here and can easily be extended to finite spin polarization and, most notably, can also be applied in higher dimensions.

1. Motivation

In recent years, tremendous effort was put forward to investigate the physics of ultracold Fermi gases, which led to the observation of a rich variety of phenomena ranging from BCS superfluidity to Bose-Einstein condensation [1, 2]. Experimentally, this was made possible by elaborate trapping and cooling techniques as well as by exploiting Feshbach resonances to influence the interaction between particles. Moreover, it became possible to realize experimental setups for low-dimensional quantum gases such that observables in two- and even one-dimensional (1D) systems are now accessible.

This variety of experimentally accessible systems calls for a corresponding theoretical investigation. Unfortunately, the resulting many-body problem is extremely challenging to solve and exact analytical solutions are either unavailable or are limited to very special cases, such as specific models in 1D. Thus, theoretical progress is mostly achieved by numerical methods. Among the most successful methods, in particular for systems beyond the few-body regime, are Quantum Monte Carlo (QMC) approaches. A major drawback, however, is given by the infamous sign-problem, for which a general solution is highly unlikely [3].

In this contribution we seek to address this problem for non-relativistic fermions in the ground state by exploiting the idea of complex stochastic quantization [4]. Although the approach has been known for some time, its rigorous mathematical foundation was unclear at first and applications remained scarce. With revived interest in recent years in the lattice QCD community (see e.g. Ref. [5]), the limitations of this approach have been formulated more rigorously and methods have been developed to mitigate them. Motivated by these advances, we present an approach to ultracold Fermi gases that relies on the Complex Langevin (CL) algorithm. This

IOP Conf. Series: Journal of Physics: Conf. Series **1041** (2018) 012006 doi:10.1088/1742-6596/1041/1/012006

allows us to extract results in a fully non-perturbative fashion as we demonstrate here for mass-imbalanced fermions in one spatial dimension with attractive and repulsive interactions. The approach is not restricted to 1D calculations but can be applied to systems of any dimension, see also Ref. [6] for a first status report on the application of this CL approach to the three-dimensional unitary spin-imbalanced Fermi gas.

2. Path integral for fermions and the sign problem

In the realm of ultracold dilute Fermi gases, the physics is dominated by s-wave scattering, i.e. contact interaction between particles. In such systems, the resulting Hamilton operator for a two-component system without an external potential is of the form

$$\hat{H} = -\sum_{s=\uparrow,\downarrow}\sum_{\mathbf{x}} \hat{\psi}_s^\dagger(\mathbf{x})\,\frac{\hbar^2}{2m_s}\vec{\nabla}^2\,\hat{\psi}_s(\mathbf{x}) \;+\; g\sum_{\mathbf{x}} \hat{\psi}_\uparrow^\dagger(\mathbf{x})\,\hat{\psi}_\uparrow(\mathbf{x})\,\hat{\psi}_\downarrow^\dagger(\mathbf{x})\,\hat{\psi}_\downarrow(\mathbf{x})\,, \tag{1}$$

where $\mathbf{x} = (x_1, \cdots, x_d)$ denotes the d-dimensional vector of spatial coordinates (which we will discretize) and g implies attractive (repulsive) interaction for $g < 0$ ($g > 0$). As usual, $\hat{\psi}_s(\mathbf{x})$ and $\hat{\psi}_s^\dagger(\mathbf{x})$ represent annihilation and creation operators for a particle of flavor s at the coordinate $\mathbf{x}$, respectively. While the first term, corresponding to the kinetic energy, is rather straightforward to deal with, the second term, i.e. the interaction part of the Hamiltonian, renders the problem difficult to solve due to its two-body nature. Although exact diagonalization is in principle possible, such a solution scales very poorly with the system size, such that other methods are needed to deal with systems beyond the few-body level.

To this end, we employ a projective method to filter out the excited states and obtain the ground state starting from an initial trial state $|\psi_T\rangle$, which we take to be a Slater determinant:

$$\mathcal{Z}_\beta = \langle\psi_T|\, e^{-\beta\hat{H}}\, |\psi_T\rangle \equiv \langle\psi_T|\, \hat{\mathcal{U}}_\beta\, |\psi_T\rangle, \tag{2}$$

The true ground state is reached in the limit of large imaginary time β, i.e. $\beta \to \infty$. In order to compute the evolution operator $\hat{\mathcal{U}}_\beta$, we first discretize imaginary time with a Trotter-Suzuki decomposition and subsequently perform a Hubbard-Stratonovich (HS) transformation to reduce the two-body operator in Eq. (1) into a sum over one-body operators coupled to an external (auxiliary) field [7, 8]. All contributions from this high-dimensional sum can be collected into a path integral which yields the partition function $\mathcal{Z}$ and is of the form

$$\mathcal{Z} \;=\; \int \mathcal{D}\sigma\ \det U_\sigma^{(\uparrow)}\ \det U_\sigma^{(\downarrow)} \;\equiv\; \int \mathcal{D}\sigma\ e^{-S[\sigma]}, \tag{3}$$

where $U_\sigma^{(s)}$ denotes the imaginary time evolution operator of the spin flavor s in single-particle representation. It can easily be appreciated from the above expression that the integrand is in general not positive semidefinite or in other words, the action S_σ is in general a complex quantity. The complex nature of the action restricts any probabilistic approach (such as importance sampling with the Metropolis algorithm) to cases where its imaginary part vanishes, i.e. to systems with equally populated spin species of equal mass and attractive pairwise interaction. In any other case such an algorithm exhibits a sign-problem and thus, the computational efforts grow exponentially with the system size. This is this reason that also restricts simulations of the repulsive Hubbard model to the half-filled case (see, e.g., Ref. [9]) as well as lattice QCD to zero baryon chemical potential (see, e.g., Ref. [10, 11]). In the following, we introduce an approach to surmount this issue by means of stochastic quantization.

IOP Conf. Series: Journal of Physics: Conf. Series **1041** (2018) 012006 doi:10.1088/1742-6596/1041/1/012006

3. Complex Langevin approach

The main idea of stochastic quantization is to use a random process, governed by the Langevin equation, to generate a probability measure $\sim e^{-S}$ that corresponds to the one in the Euclidean path integral in Eq. (3). More specifically, one introduces a fictitious Langevin time t_{CL} and evolves the auxiliary field σ according to

$$\dot{\sigma}(t) = -\frac{\partial S[\sigma(t)]}{\partial \sigma(t)} + \eta(t), \tag{4}$$

where the dot represents a derivative with respect to t_{CL} and $\eta(t)$ is a random noise term with $\langle \eta(t) \rangle = 0$ and $\langle \eta(t)\,\eta(t') \rangle = 2\delta_{t,t'}$. Note that there is one component of the auxiliary field for every spacetime lattice site. In the limit $t_{CL} \to \infty$, the distribution of σ (ideally) becomes stationary and follows $\sim e^{-S[\sigma]}$.

While this elegant approach works well for real S, complex S requires complexifying the auxiliary field and more care. Indeed, the Langevin equation acquires an imaginary part and, after discretizing t_{CL}, the evolution equations become

$$\begin{aligned} \delta\sigma_R &= -\mathrm{Re}\left[\frac{\partial S[\sigma]}{\partial \sigma}\right] h_t + \eta\sqrt{h_t}\,, \\ \delta\sigma_I &= -\mathrm{Im}\left[\frac{\partial S[\sigma]}{\partial \sigma}\right] h_t, \end{aligned} \tag{5}$$

where h_t serves as the integration time step. Although similar in structure to the real Langevin equation, two main problems arise: first, the random process may not converge and furthermore, even if the a stationary distribution is approached there is no guarantee that the results are correct and therefore further checks are in general required. The former issue is of a numerical nature and can be cured by choosing the integration step h_t to be adaptive [12]. The possibly faulty convergence, on the other hand, poses a much more serious problem and was traced back to the occurrence of singularities in the drift term $\sim \frac{\partial S[\sigma]}{\partial \sigma}$ in Eq. (5), which correspond to zeroes in the fermion determinant in our approach. While it is difficult to gain insight into the pole structure of this complicated quantity before performing the actual simulation, it is possible to analyze its behavior *a posteriori* as pointed out in Ref. [13].

The HS transform chosen for this work, which is typical in hybrid Monte Carlo (HMC) simulations, is proportional to $\sin(\sigma)$ and thus bounded in the real part but unbounded in the imaginary direction. In order to avoid singularities associated with an "unboundedly escaping" σ field, we amend the CL equations with an extra term $\sim \xi$ which we call the "regulator" [14]:

$$\begin{aligned} \delta\sigma_R &= -\mathrm{Re}\left[\frac{\partial S[\sigma]}{\partial \sigma}\right] h_t - 2\xi\sigma_{\mathrm{R}} h_t + \eta\sqrt{h_t}\,, \\ \delta\sigma_I &= -\mathrm{Im}\left[\frac{\partial S[\sigma]}{\partial \sigma}\right] h_t - 2\xi\sigma_{\mathrm{I}} h_t\,. \end{aligned} \tag{6}$$

The extra term can be thought of as a damping force that keeps the magnitude of the HS field from "escaping" too far into the complex plane. Naturally this introduces a systematic error which has to be studied carefully. By performing runs at different small but finite values of ξ and then extrapolate to the limit $\xi \to 0$ one can eventually extract the results for physical observables [14, 15].

4. Case study: 1D fermions with unequal masses

Equipped with a method that is able to deal with systems that are subject to a sign problem in case of conventional QMC methods, we now turn to a first application: the ground state of 1D

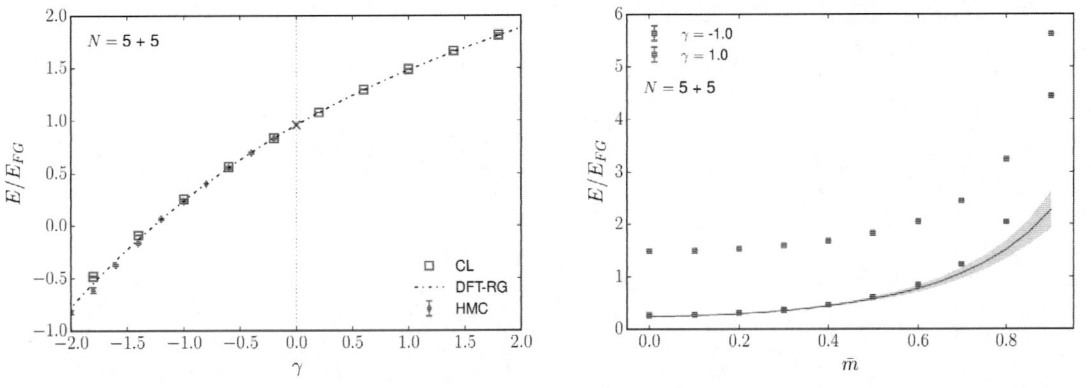

Figure 1. Ground-state energy (normalized with the non-interacting value E_{FG}) for systems of $N = 5 + 5$ particles on a lattice with $N_x = 40$ sites. Left: energies as a function of the dimensionless coupling γ. We show CL results (blue squares) alongside with HMC values for attractive couplings (red diamonds) [20] and DFT-RG results for attractive and repulsive interactions (dashed-dotted line) [21, 22]. Right: energies as a function of the mass asymmetry parameter $\bar{m}$ for couplings of $\gamma = 1.0$ and $\gamma = -1.0$. Symbols depict CL results whereas solid lines represent analytically continued iHMC results where the shaded area is the 95% confidence level associated with the corresponding fits of the data.

fermions with unequal masses. We consider Fermi mixtures of two equally populated species but finite mass difference, according to the model in Eq. (1). Note that even in 1D there is no general analytic solution to this problem and so far all attempts are restricted to specific mass/particle configurations [16, 17].

For all calculations in the following, we put a fixed number of $N = N_\uparrow + N_\downarrow$ fermions on a 1D lattice of $N_x = 40$ sites and enforce periodic boundary conditions to minimize finite-size effects. While our approach is by no means limited to such a configuration, we found it sufficient to gain a first insight and provide a benchmark for ground-state calculations in a non-relativistic setting with the CL method. In order to extrapolate to the ground state we performed simulations at large imaginary times up to ~ 1500 temporal steps with a step size of $d\tau = 0.05$ and subsequently averaged over several runs with varying temporal extent, which was found sufficient in previous studies with the HMC method [18]. Furthermore, all results reflect averages over 5000 de-correlated samples originating from a number of different random seeds, defining the initial conditions of the CL equations. In the following, we use the convention $\hbar = c = m_0 = 1$, where m_0 is the average mass across the two species. We quantify the mass-imbalance by the dimensionless mass-asymmetry parameter

$$\bar{m} = \frac{m_\uparrow - m_\downarrow}{m_\uparrow + m_\downarrow}. \tag{7}$$

In order to present all results as dimensionless quantities, we rescale energies with the non-interacting, mass-balanced energy in the continuum:

$$E_{FG} = \frac{1}{3} N \varepsilon_F = \frac{1}{6} N k_F^2 \tag{8}$$

where $k_F = \frac{\pi}{2} n$ is the Fermi wavevector. Additionally, we use the (constant) fermion density n to define the dimensionless coupling $\gamma = g/n$, as it is typical in the literature of 1D systems [19].

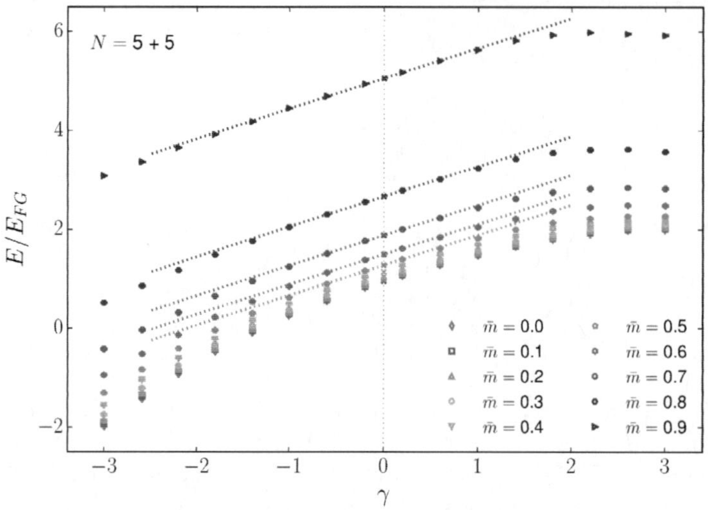

Figure 2. Ground-state energy as a function of the coupling γ for $N = 5 + 5$ fermions on a lattice of $N_x = 40$ sites for $\bar{m} = 0.0, 0.1, \ldots, 0.9$. The symbols show values from the CL approach together with values obtained from first-order lattice perturbation theory (dotted lines).

4.1. Benchmark study for balanced systems

We first consider systems with $\bar{m} = 0$, i.e. mass-balanced mixtures. In this case, we are able to compare with a variety of other methods as shown in the left panel of Fig. 1. In the attractive regime ($\gamma < 0$) our CL values agree very well with results obtained from HMC simulations, which have been shown to agree with the solutions based on the Bethe ansatz [20]. Additionally, we show results from a renormalization group approach to density functional theory (DFT-RG) and observe perfect agreement for $|\gamma| \lesssim 2.0$, which is the regime where the DFT-RG approach is currently expected to yield reliable results [21].

4.2. Feasibility study: systems with arbitrary mass asymmetry

By moving to finite $\bar{m}$, the possibilities to cross-check our result are much more limited as this regime is harder to access, especially at large imbalances. One method that is able to provide results up to medium mass imbalances is the so-called iHMC approach: by taking the mass-asymmetry to be imaginary, one obtains a positive-semidefinite measure in Eq. (3), ultimately necessary for the HMC approach. The price to pay is an analytic continuation from imaginary $\bar{m}$ to the real axis, which requires very precise expectation values in order to push the results to large $\bar{m}$. The method was successfully applied to the unitary Fermi gas [23] as well as 1D fermions with mass-imbalance in Ref. [15], to which we compare the CL results in the right panel of Fig. 1. Remarkably, the CL and iHMC results are practically indistinguishable up to $\bar{m} \sim 0.5$, which is where artefacts associated with the analytic continuation in the iHMC algorithm start to become noticeable (indicated by the large uncertainty bands). In contrast, the CL results remain smooth in this large-$\bar{m}$ regime and associated error bars stay roughly of the same magnitude as for $\bar{m} \lesssim 0.5$.

4.3. Full equation of state for 1D fermions with unequal masses

Backed up with the excellent agreement with a variety of methods, we now present the full equation of state for a large array of mass imbalances as a function of the dimensionless coupling γ in Fig. 2. The wide range of parameters studied here underscores the versatility of our CL approach as it includes regimes relevant to (future) experimental setups where quantitative results are scarce. Such experiments include Fermi mixtures of ^{6}Li and ^{40}Ka as well as ^{40}Ka and ^{161}Dy, which correspond to $\bar{m} = 0.74$ and $\bar{m} = 0.6$ in our convention, respectively.

It is necessary to mention, however, that results for repulsive interactions have to be taken with caution. For repulsive couplings we observe outliers in the distributions of the studied observables and the number of these outliers even increases with increasing coupling strength. This observation can be traced back to two different effects: First, there is a signal-to-noise issue associated with methods that use the fermion determinant as a probability measure which results in larger uncertainties than the ones associated with an assumed Gaussian distribution [24]. Second, there are implications associated with the occurrence of zeroes in the fermion determinant within the CL framework [13]. While the first issue is solvable by numerical means, albeit not implemented in this work, the second is of a conceptual nature, as mentioned above. At the present stage, however, it is difficult to separate one from the other and it is subject of ongoing investigations to resolve the origin of the observed behavior of the distributions of observables at strong repulsive couplings.

5. Conclusion & Outlook

In this contribution we have used a modified version of an algorithm known from the lattice QCD community and applied it to non-relativistic systems, allowing us to investigate areas of the parameter space in non-relativistic systems that previously have been prohibitive. As a cross-check we have performed comparisons with various methods and found remarkable agreement with our CL results. Furthermore, we computed the full equation of state for spin-balanced but mass-imbalanced fermions in one dimension, a problem that is beyond the capabilities of previously existing stochastic methods.

Although the CL approach has been known for many years and has been applied frequently in relativistic studies, it is still a method under construction, at least in non-relativistic settings. Its main benefit is the extendibility to systems at finite temperature, polarized systems as well as arbitrary dimension. Additionally, it is possible, in a straightforward manner, to compute more sophisticated observables such as density matrices and correlation functions, which enables the investigation of inhomogeneous phases, such as the Fulde-Ferell-Larkin-Ovchinnikov phase. The application of our approach to such phenomena is currently under investigation.

Acknowledgments

This work was supported by HIC for FAIR within the LOEWE program of the State of Hesse and the U.S. National Science Foundation under Grant No. PHY1452635 (Computational Physics Program). Numerical calculations have partially been performed at the LOEWE-CSC Frankfurt.

References

[1] Bloch I, Dalibard J and Zwerger W 2008 *Rev. Mod. Phys.* **3** 885
[2] Giorgini S, Pitaevskii L P and Stringari S 2008 *Rev. Mod. Phys.* **4** 1215
[3] Troyer M and Wiese U J 2005 *Phys. Rev. Lett.* **94** 170201
[4] Parisi G and Wu Y S 1981 *Sci. Sin.* **24** 483
[5] Aarts G 2009 *Phys. Rev. Lett* **102** 131601
[6] Loheac A, Braun J and Drut J E *Preprint* 1710.05020
[7] Blankenbecler R, Scalapino J D and Sugar R L 1981 *Phys. Rev. D* **24** 2278
[8] Drut J E and Nicholson A N 2013 *J. Phys. G: Nucl. Part. Phys.* **40** 043101.
[9] Hirsch J E, Phys. Rev. B **28**, 4059 (1983); Paiva T and Scalettar T and Randeria M and Trivedi N, Phys. Rev. Lett. **104**, 066406 (2010).
[10] Aarts G 2016 *J. Phys. Conf. Ser.* **706** 022004
[11] Philipsen O 2013 *Prog. Part. Nucl. Phys.* **70** 55
[12] Aarts G, James F A, Seiler E and Stamatescu I O 2011 *Eur. Phys. J. C* **71** 1756
[13] Aarts G, Seiler E, Sexty D and Stamatescu I O 2017 *JHEP* **05** 044

[14] Loheac A C and Drut J E 2017 *Phys. Rev. D* **95** 094502
[15] Rammelmüller L, Porter W J, Drut J E and Braun J 2017 *Preprint* 1708.03149
[16] Olshanii M and Jackson S G 2015 *New. Journ. Phys.* **17** 105005
[17] Harshman N L, Olshanii M, Dehkharghani A S, Volosniev A G, Jackson S G and Zinner N T *Preprint* 1704.01433
[18] Rammelmüller L, Porter W L, Drut J E and Braun J 2017 *Phys. Rev. A* **96** 033635
[19] Barth M and Zwerger W 2011 *Ann. Phys.* **326** 2544
[20] Rammelmüller L, Porter W L, Loheac A C and Drut J E 2015 *Phys. Rev. A* **92** 013631
[21] Kemler S, Pospiech M and Braun J 2017 *Journ. Phys G* **44** 015101
[22] Kemler S 2017 Ph.D. thesis, TU Darmstadt
[23] Braun J, Drut J E and Roscher D 2015 *Phys. Rev. Lett.* **114** 050404
[24] Shi H and Zhang S 2016 *Phys. Rev. E* **93** 033303

IOP Conf. Series: Journal of Physics: Conf. Series **1041** (2018) 012007 doi:10.1088/1742-6596/1041/1/012007

Nuclear Structure from the In-Medium Similarity Renormalization Group

H. Hergert[1,*], J. M. Yao[2,1], T. D. Morris[3,4], N. M. Parzuchowski[1,5], S. K. Bogner[1], J. Engel[2]

[1]NSCL/FRIB Laboratory and Department of Physics & Astronomy, Michigan State University, East Lansing, MI 48824

[2]Department of Physics & Astronomy, University of North Carolina at Chapel Hill, Chapel Hill, NC 27599

[3]Department of Physics & Astronomy, University of Tennessee-Knoxville, Knoxville, TN 37996

[4]Oak Ridge National Laboratory, Oak Ridge, TN 37830

[5]Department of Physics, The Ohio State University, Columbus, OH 43210

*Corresponding author, E-mail: hergert@frib.msu.edu

Abstract. Efforts to describe nuclear structure and dynamics from first principles have advanced significantly in recent years. Exact methods for light nuclei are now able to include continuum degrees of freedom and treat structure and reactions on the same footing, and multiple approximate, computationally efficient many-body methods have been developed that can be routinely applied for medium-mass nuclei. This has made it possible to confront modern nuclear interactions from Chiral Effective Field Theory, that are rooted in Quantum Chromodynamics with a wealth of experimental data.

Here, we discuss one of these efficient new many-body methods, the In-Medium Similarity Renormalization Group (IMSRG), and its applications in modern nuclear structure theory. The IMSRG evolves the nuclear many-body Hamiltonian in second-quantized form through continuous unitary transformations that can be implemented with polynomial computational effort. Through suitably chosen generators, we drive the matrix representation of the Hamiltonian in configuration space to specific shapes, e.g., to implement a decoupling of low- and high-energy scales, or to extract energy eigenvalues for a given nucleus.

We present selected results from Multireference IMSRG (MR-IMSRG) calculations of open-shell nuclei, as well as proof-of-principle applications for intrinsically deformed medium-mass nuclei. We discuss the successes and prospects of merging the (MR-)IMSRG with many-body methods ranging from Configuration Interaction to the Density Matrix Renormalization Group, with the goal of achieving an efficient simultaneous description of dynamic and static correlations in atomic nuclei.

1. Introduction

Effective Field Theory (EFT) and Renormalization Group (RG) methods have become important tools of modern (nuclear) many-body theory. In the Standard Model, Quantum Chromodynamics (QCD) is the fundamental theory of the strong interaction, but the description of nuclear observables on the level of quarks and gluons is not feasible, except in the lightest few-nucleon systems (see, e.g., [1]). Nowadays, nuclear interactions are derived in chiral EFT, which provides a constructive framework and organizational hierarchy not only for nuclear interactions, but also for the corresponding electroweak currents (see, e.g., [2–14]). Chiral EFT

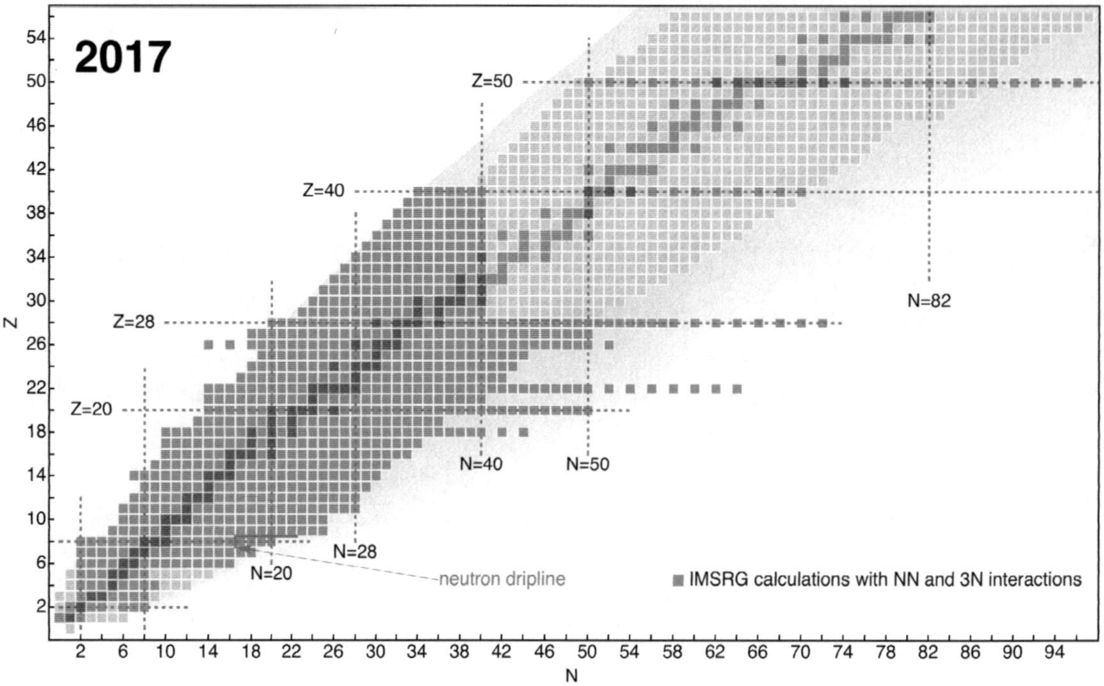

Figure 1. Survey of IMRSG-based calculations of nuclear ground-state properties and low-lying spectra up to 2017.

is essentially a low-momentum expansion of nuclear observables, whose high-momentum (short-range) physics is not explicitly resolved, but parametrized by the so-called low-energy constants (LECs) of the theory. In principle, LECs can be determined by matching results of chiral EFT and (Lattice) QCD for observables accessible by both theories, but such efforts are still in their infancy [15, 16]. In practice, LECs are therefore fit to experimental data for low-energy QCD observables, typically in the pion-nucleon (πN), two-nucleon (NN), and three-nucleon (3N) sectors [2, 3, 17, 18].

RG methods can be used to smoothly connect EFTs with different resolution scales and degrees of freedom [19, 20]. Since they were introduced in low-energy nuclear physics around the start of the millennium [21–24], they have greatly enhanced our understanding of the nuclear many-body problem by providing a systematic framework for ideas that had been discussed in the nuclear structure community since the 1950s. The Similarity Renormalization Group (SRG) [25, 26], in particular, is now widely used to decouple low- and high-momentum physics in nuclear interactions [23, 27–29]. This decoupling leads to a greatly improved convergence behavior in many-body methods that rely on a configuration space, enabling applications of these methods to increasingly heavy nuclei [29–37]. As an example, Fig. 1 shows the nuclei for which converged ground-state calculations have been performed in the In-Medium SRG (IMSRG) framework that is the focus of this work, starting from NN and 3N interactions derived in chiral EFT. This enhanced reach of *ab initio* nuclear structure theory has made it possible to confront chiral interactions with a wealth of existing nuclear data, with the goal of improving these interactions in order to reduce discrepancies between theory and experiment.

In the IMSRG, we apply the idea of decoupling energy scales to directly tackle the nuclear many-body problem [37–40]. In essence, we use RG flow equations to decouple physics at different excitation energy scales of the nucleus, and render the Hamiltonian matrix in

IOP Conf. Series: Journal of Physics: Conf. Series **1041** (2018) 012007 doi:10.1088/1742-6596/1041/1/012007

configuration space block or band diagonal in the process. With an appropriately chosen decoupling strategy, it is even possible to extract specific eigenvalues and eigenstates of the Hamiltonian, and thereby solve the quantum many-body problem. We will see that this can be achieved on the operator level, without actually constructing the Hamiltonian matrix in a factorially growing configuration space basis. We will also discuss how the IMSRG evolution can be viewed as a re-organization of the many-body expansion in which correlations are absorbed into an *RG-improved* Hamiltonian. The idea of using Hamiltonian flow equations to solve quantum many-body problems was already discussed in Wegner's initial SRG work [26] (also see [41] and references therein), and it is used in a variety of flavors in domains like solid-state physics [42–46], or quantum chemistry [47–52].

In the following, we will briefly discuss the salient elements of the chiral EFT description of the strong interaction at low energies (Sec. 2), and the use of SRG evolutions to dial the resolution scale of nuclear interactions (Sec. 3). We will then describe the IMSRG formalism and its extension to correlated reference states, with the specific goal of tackling open-shell systems with multi-reference character (Sec. 4). Selected applications of this Multireference IMSRG (MR-IMSRG) are presented in Sec. 5, followed by a discussion of how the MR-IMSRG can help us tackle collective correlations in nuclei (Sec. 6). Finally, we will touch upon the use of IMSRG-improved Hamiltonians in other many-body methods, and possible interfaces with White's Density Matrix Renormalization Group (DMRG) [53], as well as tensor RG methods for many-body systems on spatial lattices [54–56] (Sec. 7).

2. Effective Field Theories of the Strong Interaction

Quantum Chromodynamics (QCD) is the fundamental theory of the strong interaction between quarks and gluons. One of its essential features is that the strength of the interaction increases in the low-energy domain that is most relevant for the structure and reactions of atomic nuclei [57, 58]. This makes the description of all but the lightest nuclei at the QCD level inefficient at best, and impossible at worst. However, strongly interacting matter undergoes a phase transition from a quark-gluon plasma to a hadronic phase in which quarks are confined in composite particles like nucleons and pions. These particles can be used as building blocks for an EFT of the strong interaction.

Following Weinberg [59, 60], the dynamics of an EFT are governed by a Lagrangian consisting of all interactions that are consistent with the symmetries of the underlying theory. Of particular importance is the chiral symmetry of QCD: massless QCD is invariant under independent rotations of left- and right-handed quarks (hence the name chiral symmetry), expressed by the symmetry group $SU(2)_L \times SU(2)_R$ [3, 59, 61]. This symmetry is explicitly broken by the quark mass terms, but since the masses of the up and down quarks are small, it is still very relevant for the interactions of nucleons and pions, which are built from those light quarks. Since the symmetry group $SU(2)_L \times SU(2)_R$ can be equivalently represented as a product of vector and axial vector rotations, $SU(2)_V \times SU(2)_A$, chiral symmetry would imply the existence of parity doublets of strongly interacting particles which are not found in nature. Thus, we can conclude that chiral symmetry must be broken spontaneously, and it turns out that it is the axial component of the theory that undergoes this symmetry breaking. The invariance of nuclear and nucleon-pion interactions under the remaining $SU(2)_V$ group is the well-known nuclear isospin symmetry [62]. The relevant particle multiplets for chiral EFT are the proton and neutron, as well as the π_+, π_0, π_- mesons. The spontaneous breaking of chiral symmetry implies the existence of massless Goldstone bosons [63], which can be readily identified with the pions.

Another important consequence of chiral symmetry and its breaking is that only derivatives of the pion fields couple to the nucleon. Thus, interaction vertices (and Feynman diagrams) become proportional to (q/Λ_χ), where $q \sim m_\pi \sim k_F \approx 140$ MeV$/c$ is the typical momentum of a system of nucleons and pions, and Λ_χ is the so-called breakdown scale of the EFT. The breakdown scale

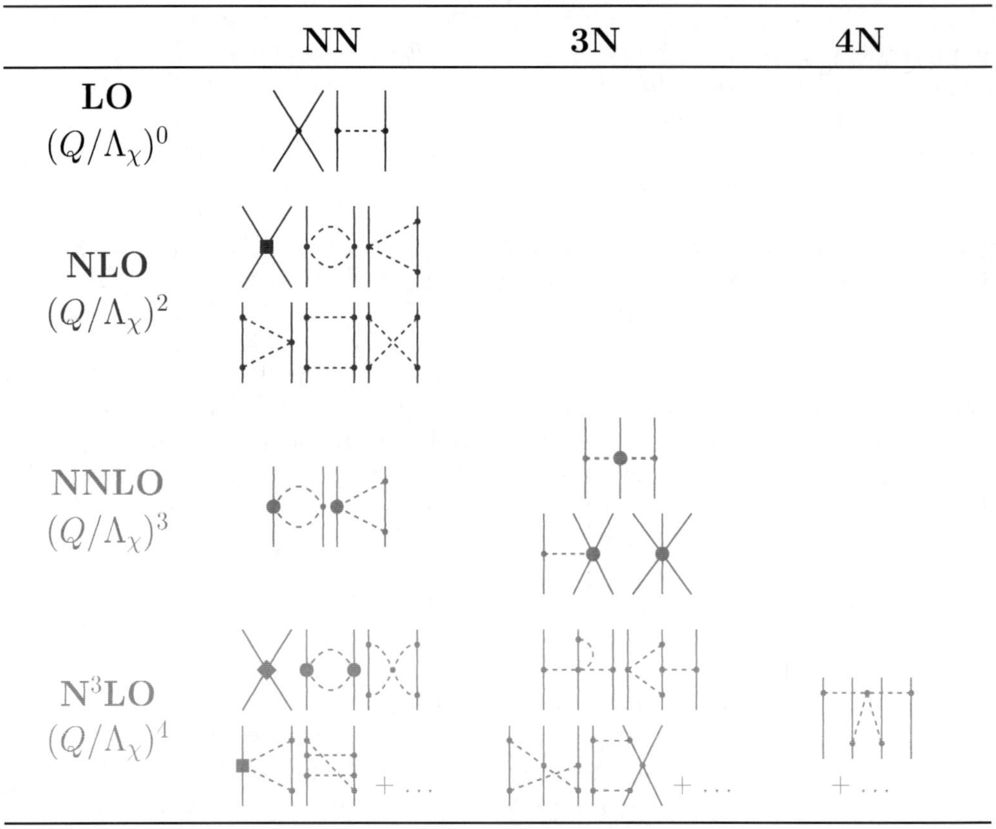

Figure 2. Chiral two-, three- and four nucleon forces through next-to-next-to-next-to-leading order (N^{3}LO) (see, e.g., [2, 3, 6]). Dashed lines represent pion exchanges between nucleons. The large solid circles, boxes and diamonds represent vertices that are proportional to low-energy constants (LECs) of the theory (see text).

is associated with physics that is not explicitly taken into account by the constructed EFT. In chiral EFT, Λ_χ is traditionally considered to be in the range 700 − 1000 MeV, with the lower value corresponding to the mass of the ρ meson that is not an explicit degree of freedom of the theory. Assumptions about Λ_χ have come under increasing scrutiny as applications in medium-mass nuclei have revealed shortcomings of the chiral interactions (see, e.g., [64] and references therein).

Nevertheless, the proportionality of the interaction vertices to $q/\Lambda_\chi < 1$ makes it possible to organize the chiral Lagrangian in powers of this ratio. Based on this power-counting scheme, one can then develop a systematic low-momentum expansion of nuclear interactions, as shown in Fig. 2 (see [2, 3, 6, 59]). These interactions consist of pion exchanges between nucleons, indicated by dashed lines, as well as nucleon contact interactions. Importantly, the diagrams shown in Fig. 2 contain different types of pion-nucleon, nucleon-nucleon, and three-nucleon vertices that are proportional to specific LECs of chiral EFT. As mentioned above, these LECs encode physics that is not explicitly resolved, either because it corresponds to a high momentum scale, or involves degrees of freedom that are not explicitly treated by the theory. Eventually, one hopes to calculate these LECs directly from the underlying QCD, i.e., through matching results from the nuclear sector to Lattice QCD calculations (see [15, 16] for first attempts in that direction). In the mean time, the LECs are fit to experimental data.

The organization of nuclear forces from leading (LO) to next-to-next-to-next-to-leading order (N^3LO) and beyond in the chiral expansion is appealing for a number of reasons. By starting from a chiral Lagrangian, we obtain a *consistent* set of two-, three- and higher many-nucleon interactions, and the power counting explains the empirical hierarchy of these nuclear forces, i.e., $V_{\text{NN}} > V_{3\text{N}} > V_{4\text{N}} > \ldots$. Moreover, the chiral Lagrangian can be readily extended with couplings to the electroweak sector by replacing the derivatives with their gauge covariant versions. This implies that the same LECs appear in the nuclear interactions and electroweak currents, and the LECs can therefore be constrained by electroweak observables (see, e.g., [11, 12, 14, 65]). Last but not least, by truncating the chiral expansion at a given order, we will eventually be able to quantify the theoretical uncertainties of the interactions and operators that serve as input for nuclear many-body theory, once some persistent issues are resolved (see, e.g., [6, 8, 9, 66–68]).

3. The Similarity Renormalization Group

As mentioned in the introduction, RG methods are natural companions for EFTs, because they make it possible to smoothly dial the resolution scale of those theories. In low-energy nuclear physics, RG methods have played an important role in formalizing ideas on the renormalization of nuclear interactions and many-body effects that had been discussed by the community for a long time (see [21–24] and references therein). For instance, local NN interactions with repulsive cores at short inter-particle distances or non-local interactions that reproduce NN scattering data equally well, can yield significantly different properties for nuclei and nuclear matter (see, e.g., [69, 70]). Colloquially, interactions with strong short-range repulsion are referred to as "hard" interactions because they induce strong many-body correlations that make the nuclear many-body problem non-perturbative. Coester and co-workers [69] showed that nuclear interactions can be softened via unitary transformations that shift strength from the repulsive core into non-local contributions, thereby making mean-field and finite-order beyond mean-field approximations to the many-body problem viable. While their work and similar investigations explicitly recognized that such transformations lead to induced many-body forces, those forces were neglected, and the discussion very much revolved around using these techniques to pin down the "true" nucleon-nucleon potential.

From the modern RG perspective, there is no such thing as a true potential, but rather infinitely many unitarily equivalent representations of low-energy QCD. In this context, the use of soft or hard NN interactions is related to the choice of a low or high resolution scale for the description of the nuclear many-body system. When an RG is used to dial the resolution, the induced many-body forces *must* be accounted for so that observables remain invariant under the RG flow (see below). For instance, induced 3N forces play a crucial role in obtaining proper nuclear matter saturation properties if we work at low resolution scale (see, e.g., [71–73]). Of course, 3N (and higher) forces are present anyway in an EFT treatment of the strong interaction, as discussed in the previous section.

3.1. SRG Evolution of Nuclear Forces

In Fig. 3 we show features of the central and tensor forces of the Argonne V18 (AV18) interaction [74] in the two-body spin-isospin channel $(S,T) = (1,0)$, which contains the deuteron bound state. AV18 is a so-called realistic interaction that describes NN scattering data with high accuracy, but precedes the modern chiral forces discussed above. It is specifically tailored to be as local as possible, to make it a suitable input for Quantum Monte Carlo calculations [7–9, 75]. This is the reason why AV18 has a strong repulsive core in the central part of the interaction. Like all NN interactions, it also has a tensor interaction resulting from pion exchange that mixes

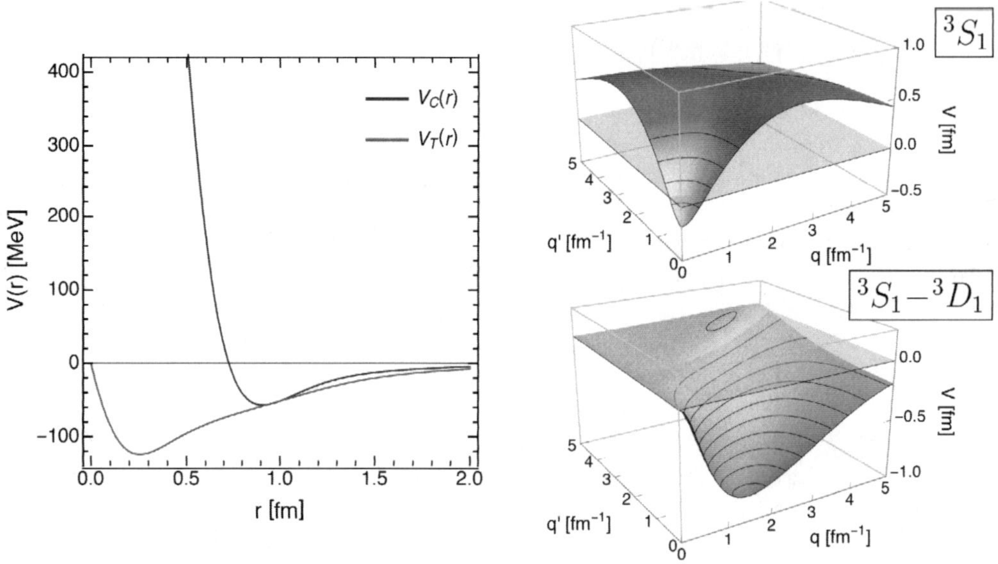

Figure 3. Repulsive core and tensor force of the Argonne V18 NN interaction [74] in the $(S, T) = (1, 0)$ channel. In the left panel, the radial dependencies of the central ($V_C(r)$) and tensor components ($V_T(r)$) of Argonne V18 are shown, while the right panel shows its momentum space matrix elements in the deuteron partial waves.

states with different orbital angular momenta,

$$V_T = v_T(r)S_{12}(r) \equiv \frac{v_T(r)}{r^2}\left(3(\vec{\sigma}_1 \cdot \vec{r})(\vec{\sigma}_2 \cdot \vec{r}) - (\vec{\sigma}_1 \cdot \vec{\sigma}_2)r^2\right), \tag{1}$$

The radial dependencies of AV18's central and tensor interactions are shown in the left panel of Fig. 3. In the corresponding momentum representation, the $^3S_1$ partial wave[1] has strong off-diagonal matrix elements, with tails extending to relative momenta as high as $|\vec{q}| \sim 20\ \text{fm}^{-1}$. The matrix elements of the $^3S_1 - ^3D_1$ mixed partial wave, which are generated exclusively by the tensor force, are sizable as well. The strong coupling between states with low and high relative momenta forces us to use large Hilbert spaces in few- and many-body calculations, even if we are only interested in the lowest eigenstates. Consequently, the eigenvalues and eigenstates of the nuclear Hamiltonian converge very slowly with respect to the basis size of the Hilbert space (see, e.g., [30]). To solve this problem, we can perform an RG evolution of the NN interaction.

First systematic applications of RG methods for the renormalization of nuclear interactions were based on RG decimation, i.e., the *projection* of the Hamiltonian onto a low-momentum (low-energy) subspace (see [21, 23] and references therein). In recent years, however, the so-called Similarity Renormalization Group (SRG) has become the RG method of choice in nuclear physics [23, 25, 26]. The SRG is based on *unitary transformations* rather than projection methods, which greatly facilitates the construction of observables that are consistent with the renormalized interaction.

The basic concept of the SRG is more general than required by RG theory: We aim to simplify the structure of the Hamiltonian in a suitable representation through a continuous

[1] We use the conventional partial wave notation $^{2S+1}L_J$, where $L = 0, 1, 2, \ldots$ is indicated by the letters $S, P, D, \ldots$. The isospin channel is fixed by requiring the antisymmetry of the NN wave function, leading to the condition $(-1)^{L+S+T} = -1$.

unitary transformation,

$$H(s) = U(s)H(0)U^\dagger(s)\,. \tag{2}$$

Here, $H(s=0)$ is the starting Hamiltonian, and the flow parameter s parameterizes the unitary transformation. To implement this transformation, we take the derivative of Eq. (2) to obtain the operator flow equation

$$\frac{d}{ds}H(s) = [\eta(s), H(s)]\,, \tag{3}$$

where the anti-Hermitian generator $\eta(s)$ is related to $U(s)$ by

$$\eta(s) = \frac{dU(s)}{ds}U^\dagger(s) = -\eta^\dagger(s)\,. \tag{4}$$

We can choose $\eta(s)$ to transform the Hamiltonian to almost arbitrary structures as we integrate the flow equation (3) for $s \to \infty$. Wegner [26] proposed the generator

$$\eta(s) \equiv [H_d(s), H_{od}(s)]\,, \tag{5}$$

which is constructed by splitting the Hamiltonian into suitably chosen *diagonal* ($H_d(s)$) and *off-diagonal* ($H_{od}(s)$) parts. The generator (5) will then suppress $H_{od}(s)$ as the Hamiltonian is evolved via equation (3) (see [23, 26, 37, 40, 41]). Here, the label diagonal merely refers to a desired structure that the Hamiltonian will assume in the limit $s \to \infty$, and not strict diagonality. We can make contact with renormalization ideas by imposing a block or band-diagonal structure on the Hamiltonian that implies a decoupling of momentum or energy scales in the RG sense.

As an example of a realistic application, Fig. 4 shows the SRG evolution of a chiral N^{3}LO NN interaction. We use a Wegner-type generator built from the relative kinetic energy in the two-nucleon system to drive the Hamiltonian towards a band-diagonal structure in momentum space:

$$\eta(\lambda) \equiv \left[\frac{\vec{q}^2}{2\mu}, v(\lambda)\right]. \tag{6}$$

We have changed variables from the flow parameter s to $\lambda = s^{-1/4}$, which has the dimensions of a momentum (in natural units), for reasons that will become clear shortly.

We start from the matrix elements of the initial interaction in the $^3S_1$ partial wave that are shown in the top row. The interaction has no appreciable strength in states with momentum differences $|\Delta\vec{q}| = |\vec{q}' - \vec{q}| \gtrsim 4.5\ \mathrm{fm}^{-1}$. By evolving the initial interaction to $\lambda = 3\ \mathrm{fm}^{-1}$ and then to $2\ \mathrm{fm}^{-1}$, the off-diagonal matrix elements are suppressed, and the interaction is almost entirely contained in a block of states with $|\Delta\vec{q}| \sim 2\ \mathrm{fm}^{-1}$, except for a weak diagonal ridge. Thus, we see that λ limits the momentum that the interaction can transfer between pairs of nucleons, and we identify λ with the resolution scale of the evolved interaction.

Next to the matrix elements, we show deuteron wave functions obtained by solving the Schrödinger equation with Hamiltonian at each λ. For the initial interaction, the $S-$wave ($L=0$) component of the wave function is suppressed at small relative distances, which reflects short-range correlations between the nucleons. There is also a significant $D-$wave ($L=2$) admixture due to the tensor interaction. As we lower the resolution scale, the "correlation hole" in the wave function is eliminated, and the $D-$wave admixture is reduced because the evolution suppresses the $^3S_1 - {}^3D_1$ matrix elements that cause this mixing [23]. The structure of the wave function becomes very simple, reminiscent of a Gaussian ansatz for a pair of independent nucleons.

IOP Conf. Series: Journal of Physics: Conf. Series **1041** (2018) 012007 doi:10.1088/1742-6596/1041/1/012007

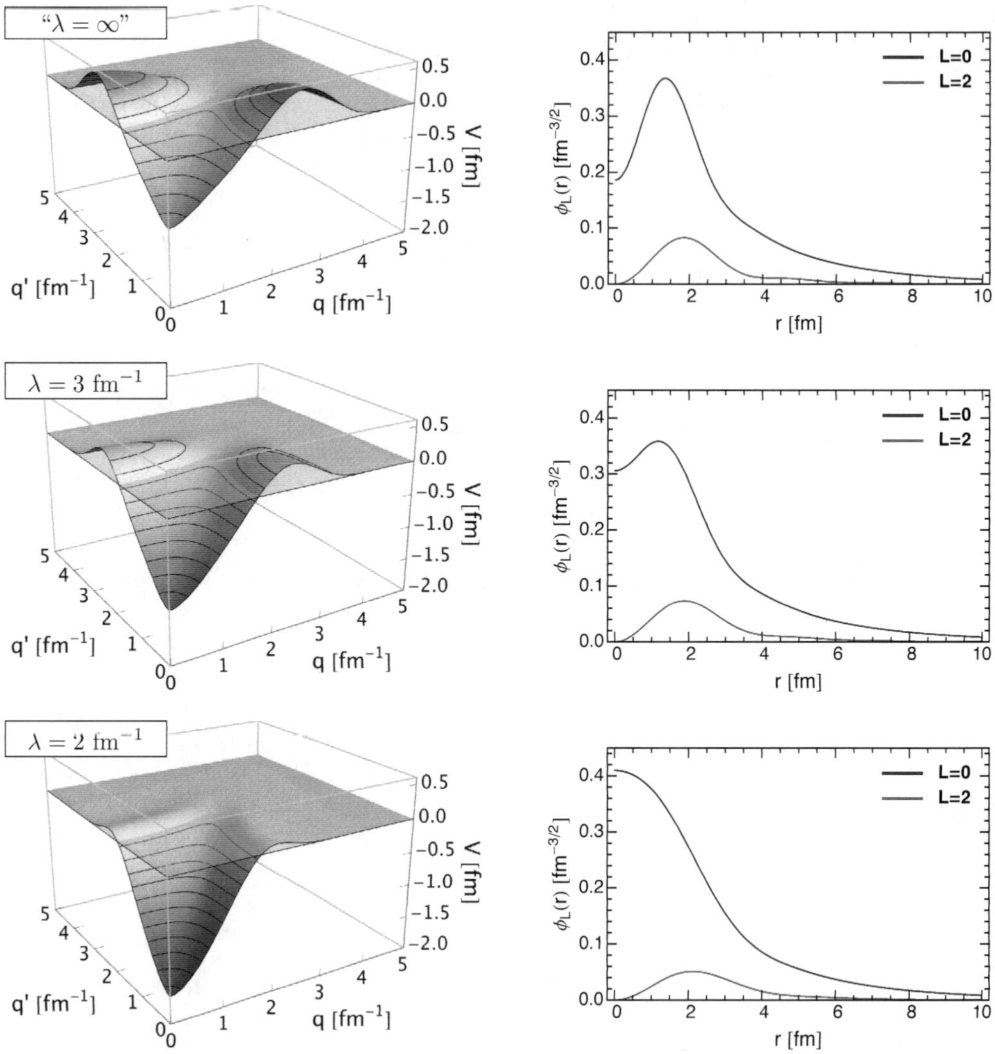

Figure 4. SRG evolution of the chiral N^{3}LO NN interaction by Entem and Machleidt [3, 76]. In the left column, we show the momentum-space matrix elements of the interaction in the $^3S_1$ partial wave for different values of the SRG resolution scale λ. The top-most row shows the initial interaction at $s = 0$ fm^4, i.e., "$\lambda = \infty$". In the right column, we show the $S-$ and $D-$wave components of the deuteron wave function that is obtained by solving the Schrödinger equation with the corresponding SRG-evolved interaction.

3.2. Induced Forces

The RG decoupling of low- and high-lying momenta prevents the Hamiltonian from scattering nucleon pairs from low- to high-momentum states, which makes it possible to converge many-body calculations in much smaller Hilbert spaces than for the harder initial interactions [30–34, 39, 40, 77–85]. However, this improvement comes at a cost, which is best illustrated by considering the Hamiltonian in second-quantized form, assuming only a two-nucleon interaction

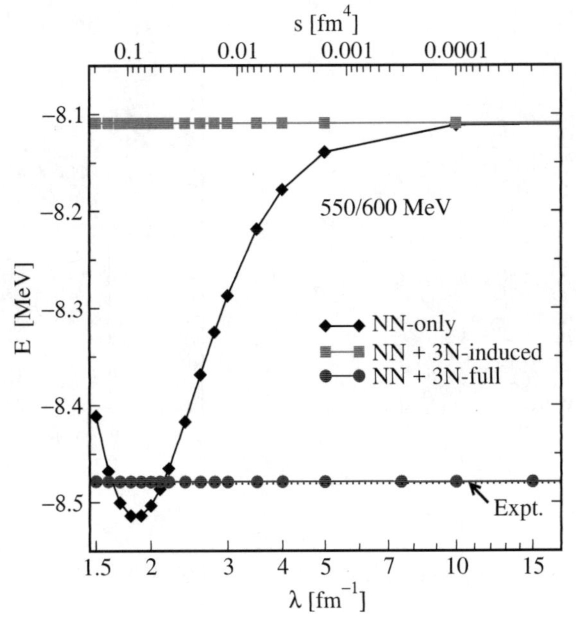

Figure 5. Ground state energy of ^{3}H as a function of the flow parameter λ for chiral NNLO NN and $NN+3N$ interactions (see [86] for details). NN-only means initial and induced $3N$ interactions are discarded, $NN+3N$-induced takes only induced $3N$ interactions into account, and $3N$-full contains initial $3N$ interactions as well. The black dotted line shows the experimental binding energy [87]. Data for the figure courtesy of K. Hebeler.

for simplicity:

$$H_{\rm int} = T_{\rm rel} + V = \frac{1}{4}\sum_{pqrs}\langle pq|\,\frac{\vec{q}_{12}^{\,2}}{2\mu} + v_{12}\,|rs\rangle a^\dagger_p a^\dagger_q a_s a_r\,. \tag{7}$$

Inserting $T_{\rm rel}$ and V into Eqs. (6) and (3), we obtain

$$[a^\dagger_i a^\dagger_j a_l a_k, a^\dagger_p a^\dagger_q a_s a_r] = \delta_{lp} a^\dagger_i a^\dagger_j a^\dagger_q a_s a_r a_k + \{a^\dagger a^\dagger a^\dagger aaa\} - \delta_{lp}\delta_{kq} a^\dagger_i a^\dagger_j a_s a_r + \{a^\dagger a^\dagger aa\}\,, \tag{8}$$

where the terms in brackets schematically stand for additional two- and three-body operators. Thus, even if we start from a pure two-body interaction, the SRG flow will induce operators of higher rank.

Nowadays, state-of-the-art SRG evolutions of nuclear interactions are performed in the three-body system [31, 86, 88–90]. In Fig. 5, we show ^{3}H ground-state energies that have been calculated with a family of SRG-evolved interactions that is generated from chiral NN and NN+3N interactions (see [86] for full details). If we truncate the evolved interaction and the SRG generator at the two-body level (*NN-only*), the SRG evolution is not unitary in the 3N system and the energy becomes λ dependent. If we truncate the operators at the three-body level instead, induced 3N interactions are properly included and unitarity is restored (*NN+3N-induced*): The energy does not change as λ is varied. Finally, for the curve *NN+3N-full* a 3N force is included in the initial Hamiltonian and evolved consistently. The resulting triton ground-state energy is invariant under the SRG flow, and it closely reproduces the experimental value because the 3N interaction's LECs are fit accordingly [2, 3, 65].

The triton example shows that it is important to track induced interactions, especially if we want to use evolved nuclear Hamiltonians as input for medium-mass or heavy nuclei

IOP Conf. Series: Journal of Physics: Conf. Series **1041** (2018) 012007 doi:10.1088/1742-6596/1041/1/012007

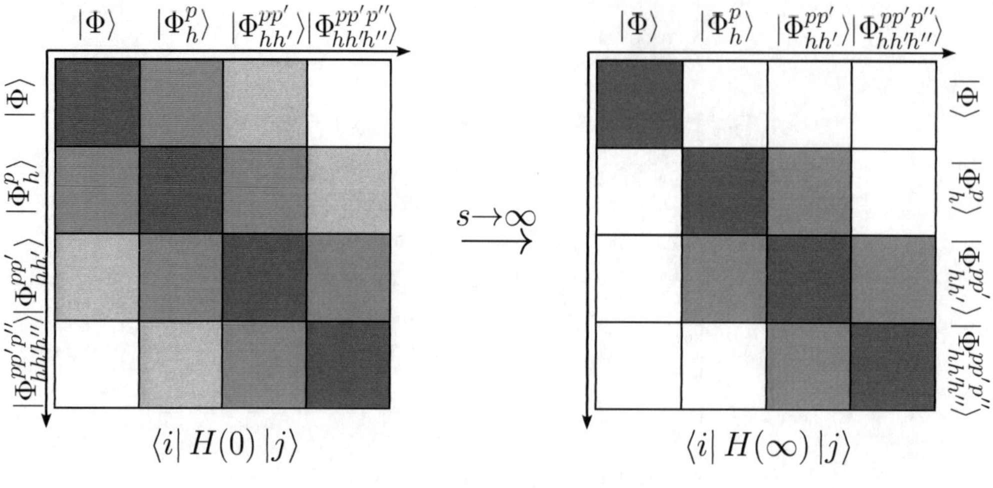

Figure 6. Schematic view of IMSRG decoupling in a many-body Hilbert space spanned by a Slater determinant reference $|\Phi\rangle$ and its particle-hole excitations $|\Phi^{p\dots}_{h\dots}\rangle$.

rather than the few-nucleon system considered so far. The nature of the SRG works at least somewhat in our favor, because truncations of the SRG flow equations lead to a violation of unitarity that manifests as a (residual) dependence of observables on the resolution scale λ. We can use this dependence as a tool to assess the size of missing contributions, although one has to take great care to disentangle them from the effects of many-body truncations [23, 32, 34, 39, 77, 85, 86, 88, 91]. The empirical observation that SRG evolutions to resolutions as low as $\lambda \sim 1.5\,\mathrm{fm}^{-1}$ appear to preserve the natural hierarchy of nuclear forces, i.e., $V_{\mathrm{NN}} > V_{\mathrm{3N}} > V_{\mathrm{4N}} > \ldots$, suggests that we can truncate induced forces whose contributions would be smaller than the desired accuracy of our calculations.

4. The In-Medium SRG

4.1. Concept

As discussed in Sec. 3, the concept of the SRG is actually quite general: We can use continuous unitary transformations to drive a given Hamiltonian (or generic operator) to a desired structure, defined by our choice of the diagonal and off-diagonal parts of the Hamiltonian, H_d and H_{od}. If we choose H_d to be the diagonal entries of a system's Hamiltonian matrix, the SRG would amount to solving the eigenvalue problem! If we consider applying this idea in genuine many-body systems like medium-mass or heavy nuclei, we quickly realize that an SRG evolution of the many-body Hamiltonian matrix would be a very inefficient approach for solving the Schrödinger equation, because the RHS of Eq. (3) forces us to repeatedly perform products of matrices whose dimension grows factorially with the number of particles and single-particle basis states. Moreover, we would essentially be working to obtain the complete spectrum of that large matrix representation rather than just a comparably small number of low-lying states that we are truly interested in.

Instead of this naive approach, we can evolve the Hamiltonian operator, and achieve the desired decoupling without ever constructing a large matrix. To this end, let us consider a representation of the Hamiltonian in a basis built from a reference Slater determinant $|\Phi\rangle$ and its particle-hole excitations, as shown in the left panel of Fig. 6. For simplicity, we show a two-body Hamiltonian, which can at most couple the reference state $|\Phi\rangle$ to 2p2h excitations, and in general $npnh$ to $(n\pm 2)$p$(n\pm 2)$h states, yielding the schematic band structure on display.

IOP Conf. Series: Journal of Physics: Conf. Series **1041** (2018) 012007 doi:10.1088/1742-6596/1041/1/012007

We now want to continuously evolve this Hamiltonian to the shape shown in the right panel of Fig. 6. Inspecting the evolved Hamiltonian more closely, we see that we aim to decouple the matrix element $\langle\Phi| H(s) |\Phi\rangle$ from all excitations in the limit $s \to \infty$ — if we succeed, we will have extracted an eigenvalue of H, and our unitary transformation will provide us with a mapping of the Slater determinant $|\Phi\rangle$ onto an exact eigenstate $|\Psi\rangle$ of $H(0)$, typically the ground state.

Since we are working with a reference state and its excitations, it is convenient to represent the Hamiltonian in terms of normal-ordered operators. We introduce the compact notation

$$A^{i_1\ldots i_k}_{j_1\ldots j_k} \equiv a^\dagger_{i_1}\ldots a^\dagger_{i_k} a_{j_k}\ldots a_{j_1} \tag{9}$$

for strings of creation and annihilation operators, and the normal ordered operators

$$:A^i_k: = A^i_k - \rho^i_k \tag{10}$$

$$:A^{ij}_{kl}: = A^{ij}_{kl} - :A^i_k:\rho^j_l + \text{perm.} - \rho^i_k\rho^j_l + \text{perm.} \tag{11}$$

$$\ldots$$

where $\rho^i_k = \langle\Phi| A^i_k |\Phi\rangle$ is the usual one-body density matrix of the reference state (see [37, 40] for details). An intrinsic nuclear A-body Hamiltonian with NN and 3N interactions,

$$H = T - T_{\text{cm}} + V_2 + V_3 \tag{12}$$

can then be exactly rewritten as

$$H = E + \sum_{ij} f^i_j :A^i_j: + \frac{1}{4}\sum_{ijkl}\Gamma^{ij}_{kl} :A^{ij}_{kl}: + \frac{1}{36}\sum_{ijklmn} W^{ijk}_{lmn} :A^{ijk}_{lmn}:, \tag{13}$$

with the individual zero- through three-body contributions

$$E \equiv \left(1-\frac{1}{A}\right)\sum_{ab} t^a_b\rho^a_b + \frac{1}{4}\sum_{abcd}\left(\frac{1}{A}t^{ab}_{cd} + v^{ab}_{cd}\right)\rho^{ab}_{cd} + \frac{1}{36}\sum_{abcdef} v^{abc}_{def}\rho^{abc}_{def}, \tag{14}$$

$$f^i_j \equiv \left(1-\frac{1}{A}\right)t^i_j + \sum_{ab}\left(\frac{1}{A}t^{ia}_{jb} + v^{ia}_{jb}\right)\rho^a_b + \frac{1}{4}\sum_{abcd} v^{iab}_{jcd}\rho^{ab}_{cd}, \tag{15}$$

$$\Gamma^{ij}_{kl} \equiv \frac{1}{A}t^{ij}_{kl} + v^{ij}_{kl} + \sum_{ab} v^{ija}_{klb}\rho^a_b, \tag{16}$$

$$W^{ijk}_{lmn} \equiv v^{ijk}_{lmn}. \tag{17}$$

Here, $\rho^{a\ldots}_{i\ldots} = \langle\Phi| A^{i_1\ldots i_k}_{j_1\ldots j_k} |\Phi\rangle$ denote the k-body density matrices of the reference state. For reasons that will become clear below, we refrain from simplifying these expressions by using the factorization of Slater determinant's density matrices. We note that due to the normal ordering, the zero-, one-, and two-body coefficients all contain *in-medium* contributions from the three-body interaction, i.e., contractions of three-body matrix elements with the density matrices of the reference state. This explains the name In-Medium SRG, and will play an important role when we introduce truncations of the IMSRG flow equations below.

Using this definition of the Hamiltonian, we can use Wick's theorem to evaluate the many-body matrix elements that couple $|\Phi\rangle$ to excitations, i.e.,

$$\langle\Phi^p_h| :H: |\Phi\rangle = \langle\Phi| :A^h_p::H: |\Phi\rangle = f^p_h \tag{18}$$

$$\langle\Phi^{pp'}_{hh'}| :H: |\Phi\rangle = \langle\Phi| :A^{hh'}_{pp'} :: H : |\Phi\rangle = \Gamma^{pp'}_{hh'} \tag{19}$$

(see [37, 40]). These matrix elements define our off-diagonal Hamiltonian

$$H_{od} \equiv \sum_{ph} f^p_h :A^p_h: + \frac{1}{4} \sum_{pp'hh'} \Gamma^{pp'}_{h'h} :A^{pp'}_{hh'}: + \text{H.c.}\,, \tag{20}$$

which can then be used in Eq. (5) or another suitable ansatz for the generator (see [37, 40] for details) to evolve the Hamiltonian to the desired form shown in the right panel of Fig. 6. Note that the elimination of H_{od} also removes the couplings between general $np nh$ and $(n\pm2)$p$(n\pm2)$h states, i.e., the outermost side-diagonal blocks.

4.2. Correlated Reference States

The discussion so far was based on the use of a Slater determinant as the reference state. On the one hand, such an independent-particle reference is convenient because it allows us to distinguish particle and hole states in our single-particle basis, but on the other hand, it means that the unitary transformation must encode all many-body correlations. The unitarity of the IMSRG transformation offers us more flexibility, because we can control to what extent correlations are described by either the Hamiltonian or the reference state. To see this, we consider the stationary Schrödinger equation and apply $U(s)$:

$$\left[U(s)HU^\dagger(s)\right] U(s)\,|\Psi_k\rangle = E_k U(s)\,|\Psi_k\rangle\,. \tag{21}$$

$U(s)$ shifts correlations from the wave function into the evolved, RG-improved Hamiltonian $H(s) = U(s)HU^\dagger(s)$. In many-body calculations with this Hamiltonian, the many-body method now “only” needs to approximate $U(s)\,|\Psi_k\rangle$ rather than $|\Psi_k\rangle$ — in the case discussed so far, even a Slater determinant would be sufficient because $U(s)$ or $H(s)$ describe the correlations[2]. There can be situations in which it is worthwhile to use the reference state to describe certain types of correlations, especially if it can capture them more efficiently than through the particle-hole type expansion in which the IMSRG-evolved Hamiltonian is set up (see Sec. 6).

The IMSRG framework can be extended to correlated reference states using a generalized normal ordering developed by Kutzelnigg and Mukherjee [92, 93]. It requires a new ingredient, the *irreducible k-body density matrices* $\lambda^{(k)}$. In the one-body case, this is just the usual density matrix

$$\lambda^i_k \equiv \rho^i_k = \langle\Phi|\, A^i_k\, |\Phi\rangle\,, \tag{22}$$

while for $k \geq 2$,

$$\lambda^{ij}_{kl} \equiv \rho^{ij}_{kl} - \mathcal{A}\{\lambda^i_k\lambda^j_l\}\,, \tag{23}$$

$$\lambda^{ijk}_{lmn} \equiv \rho^{ijk}_{lmn} - \mathcal{A}\{\lambda^i_l\lambda^{jk}_{mn}\} - \mathcal{A}\{\lambda^i_l\lambda^j_m\lambda^k_n\}\,, \tag{24}$$

etc., where $\mathcal{A}\{\ldots\}$ fully antisymmetrizes the indices of the expression within the brackets. The irreducible densities encode the correlation content of an arbitrary reference state $|\Phi\rangle$. If the reference state is a Slater determinant, i.e., an independent-particle state, the full two-body density matrix factorizes, and $\lambda^{(2)}$ vanishes (see, e.g., [37]).

Normal-ordered one-body operators are constructed in the same manner as in the standard normal ordering, e.g.,

$$:A^a_b: = A^a_b - \lambda^a_b\,, \tag{25}$$

[2] We also recall the evolution of the NN interaction in Sec. 3, where the solution of the Schrödinger equation for the deuteron with an SRG-evolved Hamiltonian was a very simple two-nucleon wave function which nevertheless provides the correct deuteron ground-state energy.

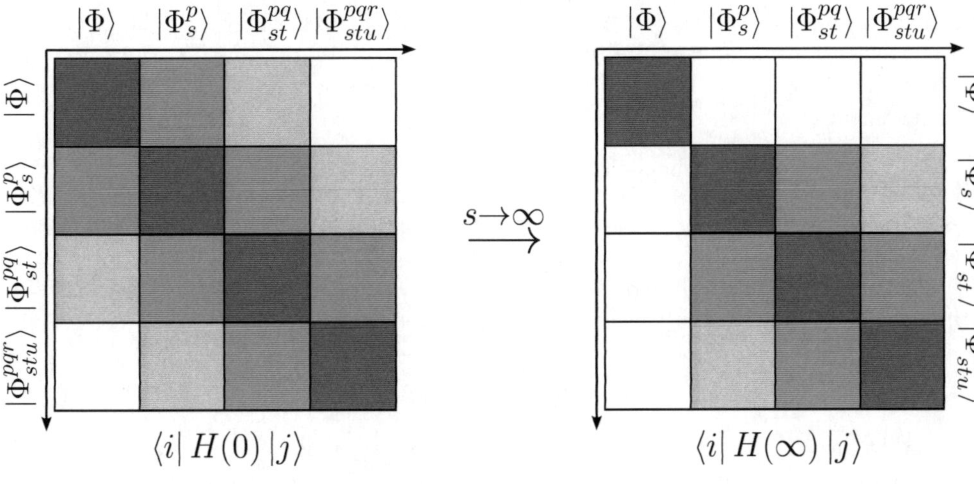

Figure 7. Schematic view of MR-IMSRG decoupling in the many-body Hilbert space. $|\Phi\rangle$ denotes an arbitrary reference state, and $|\Phi^{p\cdots}_{s\cdots}\rangle$ are suitably defined excitations.

$$:A^{ab}_{cd}: = A^{ab}_{cd} - \lambda^a_c :A^b_d: + \lambda^a_d :A^b_c: - \lambda^b_d :A^a_c: + \lambda^b_c :A^a_d: - \lambda^a_c\lambda^b_d + \lambda^a_d\lambda^b_c - \lambda^{ab}_{cd}\,, \tag{26}$$

$$\ldots$$

The expressions for the coefficients of the normal-ordered nuclear Hamiltonian, Eqs. (14)–(17) remain unchanged, because we wrote them in terms of the full density matrices. When we work with arbitrary reference states, the regular Wick's theorem extended with additional contractions like

$$:A^{\boxed{ab}}_{cd} :: A^{ij}_{\boxed{kl}} : = +\lambda^{\boxed{ab}}_{\boxed{kl}} :A^{ij}_{cd}:\,, \tag{27}$$

$$:A^{\boxed{ab}}_{c\boxed{d}} :: A^{\boxed{i}j}_{\boxed{kl}} : = -\lambda^{\boxed{abi}}_{\boxed{dkl}} :A^{j}_{c}:\,, \tag{28}$$

$$:A^{\boxed{ab}}_{\boxed{cd}} :: A^{\boxed{ij}}_{\boxed{kl}} : = +\lambda^{\boxed{abij}}_{\boxed{cdkl}}\,. \tag{29}$$

(see [37] for a detailed discussion). These contractions increase the number of terms when we expand operator products, e.g., in the analysis of many-body matrix elements or the flow equations (see below), but fortunately, the overall increase in complexity is manageable.

In Fig. 7, we illustrate the schematic decoupling for the Hamiltonian in a basis spanned by a correlated reference state and its excitations. Since our reference state is no longer a Slater determinant, there are no well-defined notions of a Fermi surface or particle and hole states any longer, and the indices of the excitation operators run over all single-particle states instead. Per construction, the reference state $|\Phi\rangle$ is still orthogonal to general excitations $:A^{i_1\ldots i_k}_{j_1\ldots j_k}:|\Phi\rangle$ because

$$\langle\Phi|:A^{i_1\ldots i_k}_{j_1\ldots j_k}:|\Phi\rangle\,. \tag{30}$$

The excitations themselves, however, are in general not orthogonal to each other, and may even be linearly dependent. As a result, the structure of the evolved MR-IMSRG Hamiltonian, shown in the right panel of Fig. 7, is not simplified as much as in the Slater determinant case, since matrix elements in the outer diagonals are not necessarily removed when novel contractions like (27)–(29) are in play.

In the MR-IMSRG, the linear dependence implies that we are implicitly transforming a rank-deficient many-body Hamiltonian matrix and are possibly working harder than we need to by

IOP Conf. Series: Journal of Physics: Conf. Series **1041** (2018) 012007 doi:10.1088/1742-6596/1041/1/012007

dragging along a space of spurious eigenstates with zero eigenvalues. Since the nuclear states we are interested in have absolute binding energies of several hundred MeVs, there has been no evidence that the presence of these spurious eigenstates adversely affects the physical states in practical calculations.

4.3. MR-IMSRG Flow Equations

We are now ready to formulate the MR-IMSRG flow equations by applying the generalized normal ordering and Wick's theorem to Eq. (3). We express all operators in terms of normal-ordered strings of creation and annihilation operators. As discussed in Sec. 3, each evaluation of the commutator in Eq. (3) induces operators of higher rank,

$$[:A^{ab}_{cd}:, :A^{ij}_{kl}:] = \delta_{ci} :A^{abj}_{dkl}: + \ldots, \tag{31}$$

hence we need to introduce truncations to obtain a computationally feasible many-body method. In contrast to Eq. (8), we are now working with *normal-ordered* operators whose in-medium contributions have been absorbed into terms of lower rank and we expect the induced operators to be much weaker than in the free-space SRG case. There is ample empirical evidence that the residual normal-ordered three-body piece can be neglected to a very good approximation in nuclear physics [37, 80, 94]. In our framework, we will eventually have to consider how the truncated terms may feed back into operators of lower rank as we evolve the flow equations (see below). In practice, we presently truncate all flowing operators at the two-body level to close the system of flow equations:

$$\eta(s) \approx \eta^{(1)}(s) + \eta^{(2)}(s)\,, \tag{32}$$

$$H(s) \approx E(s) + f(s) + \Gamma(s)\,, \tag{33}$$

$$\frac{d}{ds}H(s) \approx \frac{d}{ds}E(s) + \frac{d}{ds}f(s) + \frac{d}{ds}\Gamma(s)\,. \tag{34}$$

This is the so-called MR-IMSRG(2) truncation [32, 38–40, 78, 95, 96], which is a cousin to a variety of other truncated many-body schemes, in particular Coupled Cluster with Singles and Doubles excitations (CCSD) [35, 47–51, 97–102], although the latter is based on non-unitary similarity transformations.

The MR-IMSRG(2) flow equations are given by [32, 37, 78]

$$\begin{aligned}\frac{dE}{ds} &= \sum_{ab}(n_a - n_b)\eta^a_b f^b_a + \frac{1}{4}\sum_{abcd}\left(\eta^{ab}_{cd}\Gamma^{cd}_{ab} - \Gamma^{ab}_{cd}\eta^{cd}_{ab}\right) n_a n_b \bar{n}_c \bar{n}_d \\ &\quad + \frac{1}{4}\sum_{abcd}\left(\frac{d}{ds}\Gamma^{ab}_{cd}\right)\lambda^{ab}_{cd} + \frac{1}{4}\sum_{abcdklm}\left(\eta^{ab}_{cd}\Gamma^{kl}_{am} - \Gamma^{ab}_{cd}\eta^{kl}_{am}\right)\lambda^{bkl}_{cdm}\,,\end{aligned} \tag{35}$$

$$\begin{aligned}\frac{d}{ds}f^i_j &= \sum_a \left(\eta^i_a f^a_j - f^i_a \eta^a_j\right) + \sum_{ab}\left(\eta^a_b \Gamma^{bi}_{aj} - f^a_b \eta^{bi}_{aj}\right)(n_a - n_b) \\ &\quad + \frac{1}{2}\sum_{abc}\left(\eta^{ia}_{bc}\Gamma^{bc}_{ja} - \Gamma^{ia}_{bc}\eta^{bc}_{ja}\right)(n_a \bar{n}_b \bar{n}_c + \bar{n}_a n_b n_c) \\ &\quad + \frac{1}{4}\sum_{abcde}\left(\eta^{ia}_{bc}\Gamma^{de}_{ja} - \Gamma^{ia}_{bc}\eta^{de}_{ja}\right)\lambda^{de}_{bc} + \sum_{abcde}\left(\eta^{ia}_{bc}\Gamma^{be}_{jd} - \Gamma^{ia}_{bc}\eta^{be}_{jd}\right)\lambda^{ae}_{cd} \\ &\quad - \frac{1}{2}\sum_{abcde}\left(\eta^{ia}_{jb}\Gamma^{cd}_{ae} - \Gamma^{ia}_{jb}\eta^{cd}_{ae}\right)\lambda^{cd}_{be} + \frac{1}{2}\sum_{abcde}\left(\eta^{ia}_{jb}\Gamma^{bc}_{de} - \Gamma^{ia}_{jb}\eta^{bc}_{de}\right)\lambda^{ac}_{de}\,,\end{aligned} \tag{36}$$

IOP Conf. Series: Journal of Physics: Conf. Series **1041** (2018) 012007 doi:10.1088/1742-6596/1041/1/012007

$$
\begin{aligned}
\frac{d}{ds}\Gamma^{ij}_{kl} = &\sum_a \left(\eta^i_a\Gamma^{aj}_{kl} + \eta^j_a\Gamma^{ia}_{kl} - \eta^a_k\Gamma^{ij}_{al} - \eta^a_l\Gamma^{ij}_{ka} - f^i_a\eta^{aj}_{kl} - f^j_a\eta^{ia}_{kl} + f^a_k\eta^{ij}_{al} + f^a_l\eta^{ij}_{ka}\right) \\
&+ \frac{1}{2}\sum_{ab}\left(\eta^{ij}_{ab}\Gamma^{ab}_{kl} - \Gamma^{ij}_{ab}\eta^{ab}_{kl}\right)(1 - n_a - n_b) \\
&+ \sum_{ab}(n_a - n_b)\left(\left(\eta^{ia}_{kb}\Gamma^{jb}_{la} - \Gamma^{ia}_{kb}\eta^{jb}_{la}\right) - \left(\eta^{ja}_{kb}\Gamma^{ib}_{la} - \Gamma^{ja}_{kb}\eta^{ib}_{la}\right)\right) ,
\end{aligned}
\tag{37}
$$

where the s-dependence has been suppressed for brevity. We work in the natural orbital basis obtained by diagonalizing the one-body density matrix,

$$
\lambda^i_k = n_i\delta^i_k \,, \tag{38}
$$

where $0 \leq n_i \leq 1$ indicates the occupation of the i-th orbital in the correlated reference state. Analogously, $0 \leq \bar{n}_i = 1 - n_i \leq 1$ are the eigenvalues of the hole density matrix

$$
\xi^i_j \equiv \lambda^i_j - \delta^i_j \,. \tag{39}
$$

Note that due to the use of general reference states, the MR-IMSRG flow equations also include couplings to correlated pairs and triples of nucleons in the reference state through the irreducible density matrices $\lambda^{(2)}$ and $\lambda^{(3)}$. The single-reference limit for a Slater determinant reference state can be obtained by setting the irreducible density matrices $\lambda^{(2)}$ and $\lambda^{(3)}$ to zero, and noting that the occupation numbers now become either 0 or 1.

Superficially, the computational cost for the evaluation of the MR-IMSRG(2) flow equations is dominated by the final term of Eq. (35), which is of $\mathcal{O}(N^7)$, where N is the size of the single-particle basis. However, most practical ansätze for correlated reference states impose conditions on the density matrix that strongly limit the number of non-zero matrix elements. For instance, the particle-number projected Hartree-Fock-Bogoliubov (HFB) [103] reference states used in our initial applications (see Sec. 5) have almost diagonal $\lambda^{(k)}$ matrices. Moreover, relevant correlations are often restricted to a so-called valence or active space consisting of a small number of single-particle states, so that only density matrix elements with *all* indices belonging to this space are non-zero. Thus, the main driver of the computational effort is the integration of the two-body flow equation, at $\mathcal{O}(N^6)$, which is the same naive scaling as for CCSD and similar methods. For closed-shell references, we can exploit the distinction of particle and hole indices to achieve exactly the same scaling as for single-reference CCSD theory, namely $\mathcal{O}(N_h^2 N_p^4)$.

5. Selected Applications

5.1. Reshuffling of Ground-State Correlations: ^{40}Ca

To illustrate the effect of the (MR-)IMSRG flow in practice, we first consider a ground-state calculation for the closed-shell nucleus ^{40}Ca. As our reference, we use a Slater determinant that is optimized by performing a Hartree-Fock calculation with a chiral NN+3N interaction. Figure 8 shows the IMSRG(2) ground-state energy $E(s)$ of ^{40}Ca as a function of the flow parameter, which corresponds to the zero-body piece of the evolved normal-ordered Hamiltonian, Eq. (14). As we integrate the flow equations to increasing s, the Hamiltonian is RG-improved with correlations that are otherwise only accessible with post-Hartree Fock methods (see Refs. [37, 40] for details). This is evident if we also consider the size of the second and third-order MBPT energy corrections, evaluated with $H(s)$. After a few integration steps, the energy corrections are absorbed into the Hamiltonian, which implies that we could perform a Hartree-Fock calculation with $H(s)$ and obtain the fully correlated ground-state energy, up to truncation errors!

Figure 8 also illustrates that the size of the MBPT corrections depends significantly on the resolution scale of the Hamiltonian (see Sec. 3). At the higher resolution scale $\lambda = 2.24$ fm^{-1}, we

IOP Conf. Series: Journal of Physics: Conf. Series **1041** (2018) 012007 doi:10.1088/1742-6596/1041/1/012007

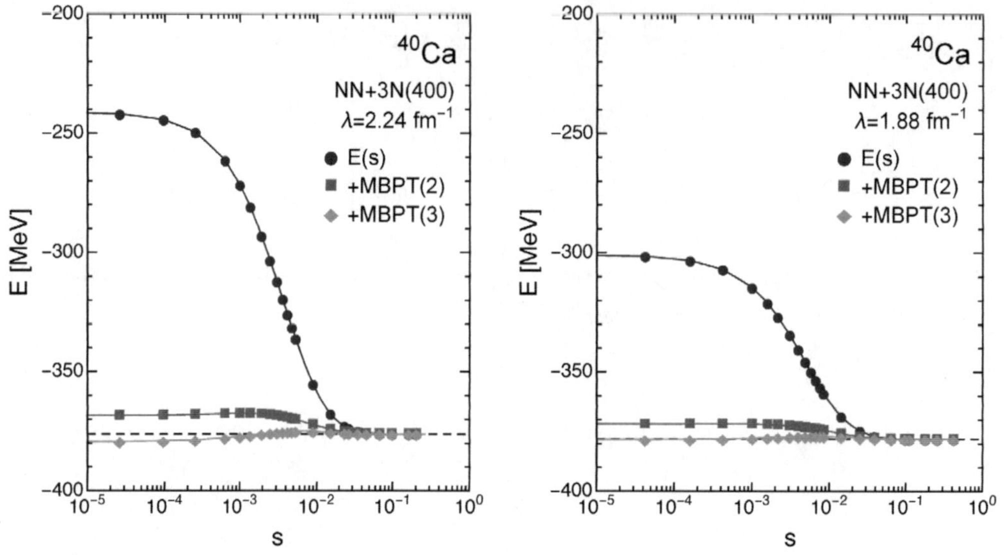

Figure 8. IMSRG(2) flow for ^{40}Ca for a chiral NN+3N interaction, denoted NN+3N(400) in the following, at two different resolution scales $\lambda = 2.24\,\mathrm{fm}^{-1}$ and $1.88\,\mathrm{fm}^{-1}$ (see Refs. [37, 40] for details). We show the flowing ground-state energy $E(s)$, and the sum of $E(s)$ and perturbative energy corrections evaluated with the evolving Hamiltonian $H(s)$ to illustrate the re-shuffling of correlations into the Hamiltonian (see text). The dashed lines indicate the final IMSRG(2) energies.

gain about 130 MeV of binding energy from correlations. At the lower resolution $\lambda = 1.88\ \mathrm{fm}^{-1}$, the HF reference state is already significantly lower in energy, so the energy gains from many-body correlations are less pronounced. This behavior is expected as interactions become increasingly soft, and thereby more perturbative (see, e.g., [23, 29]). Note that the final ground-state energies for $\lambda = 2.24\ \mathrm{fm}^{-1}$ and $1.88\ \mathrm{fm}^{-1}$ are almost identical. As discussed in section 3, all results should ideally be invariant under arbitrary changes of λ, which seems to be the case for the specific NN+3N Hamiltonian we used here [33, 77].

5.2. Ground-State Energies Along Isotopic Chains: The Oxygen Isotopes

The oxygen isotopic chain is an important testing ground for *ab initio* nuclear structure theory because it is accessible to a wide array of exact and approximate many-body methods [32, 77, 104–111]. As a bonus, the semi-magicity of the oxygen isotopes — the $Z = 8$ protons are in a closed-shell configuration — allows us to exploit spherical symmetry in our calculations. In Fig. 9, we compare MR-IMSRG(2) ground-state energies, obtained using particle-number projected HFB vacua [32, 103, 112] as references for the open-shell nuclei, with a variety of other methods:

- the Importance-Truncated No-Core Shell Model (IT-NCSM), which is an essentially exact diagonalization [30, 113],
- two types of Coupled Cluster, CCSD and Λ-CCSD(T), the latter of which includes approximate 3p3h (Triples) excitations [114, 115], and
- the ADC(3) Self-Consistent Green's Function (SCGF) approach [104, 105] (obtained at a slightly different resolution, exploiting the approximate invariance of the energies under variations of λ).

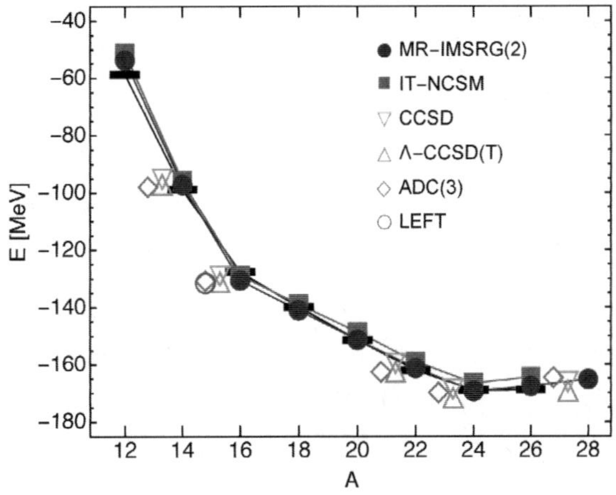

Figure 9. Ground-state energies of the oxygen isotopes from MR-IMSRG(2) and other many-body approaches, using the NN+3N(400) interaction at $\lambda = 1.88\,\text{fm}^{-1}$. Some data points were offset horizontally to enhance the readability of the figure. The ADC(3) Self-Consistent Green's Function results [104, 105] were obtained for $\lambda = 2.0\ \text{fm}^{-1}$, but the dependence of energies on λ is very weak (see Fig. 8). Black bars indicate experimental data [87].

Importantly, all used methods give consistent results when the same input Hamiltonian is used, and the theoretical ground-state energies agree within a few percent with experimental ground state energies. The systematically truncated methods, i.e., MR-IMSRG(2), CCSD, Λ-CCSD(T) and ADC(3), agree very well with the exact IT-NCSM results, on the level of 1%–2%, which provides us with an estimate of their truncation errors. The MR-IMSRG(2) ground-state energy of ^{16}O also agrees well with the result of a Nuclear Lattice EFT (NLEFT) calculation [108] directly using the chiral NNLO Lagrangian. Since the treatment of the nuclear many-body problem in NLEFT is completely different from all the other approaches compared here [116], the good agreement is very encouraging.

The *ab initio* calculations predict the neutron drip line at ^{24}O in accordance with experimental findings [117, 118]. The addition of further neutrons no longer produces bound oxygen isotopes. While absolute ground-state energies can change significantly under variations of the resolution scale or other modifications of the initial Hamiltonian, the drip line signal turns out to be rather robust [32], if somewhat exaggerated: In recent years, experimental studies have repeatedly revised the energy of the ^{26}O resonance downward [118–120]. Experimental data for ^{28}O is forthcoming, which will most likely settle the issue of the oxygen drip line. Still, the complex interplay of nuclear interactions, many-body and continuum effects causes the flat trend of experimental ground-state resonance energies beyond ^{24}O that will preserve the oxygen chain's status as an important testing ground for nuclear Hamiltonians and many-body methods for the foreseeable future.

While chiral NN+3N interactions give a good reproduction of the oxygen ground-state energies, their performance can differ significantly for other observables, as illustrated in Fig. 10. The NN+3N(400) Hamiltonian [65, 77] that was also used in Figs. 8 and 9 underestimates the experimentally known oxygen charge radii by about 10%, and misses the sharp increase for ^{18}O. The other shown interaction, NNLO$_{\text{sat}}$ [17] improves the situation significantly, in part because its LECs (cf. Sec. 2) are optimized with a protocol that includes selected many-body data, including the charge radius of ^{16}O — however, it also falls short in ^{18}O [122].

As we go to heavier nuclei, the NN+3N(400) increasingly overestimates the nuclear binding energy, up to about 1 MeV/nucleon in the light tin isotopes, which are the heaviest nuclei

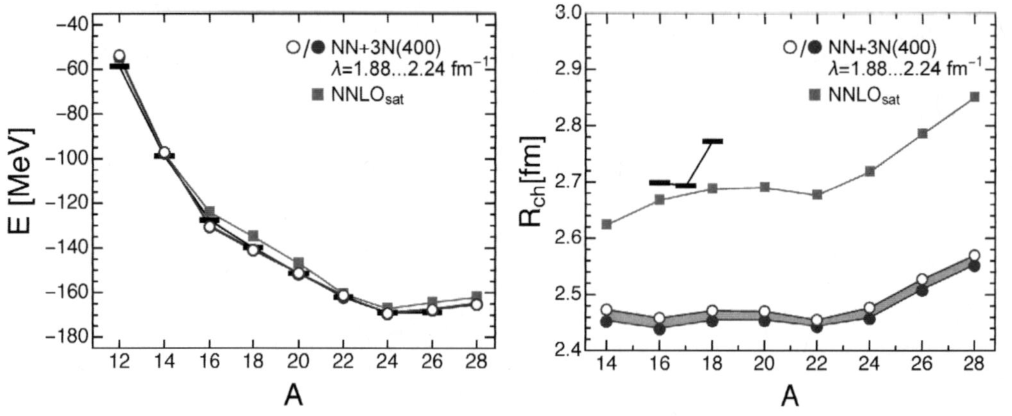

Figure 10. MR-IMSRG(2) ground-state energies and charge radii of the oxygen isotopes for NNLO_{sat} and NN+3N(400) at $\lambda = 1.88, \ldots, 2.24\,\text{fm}^{-1}$ (see text). Black bars indicate experimental data [87, 121].

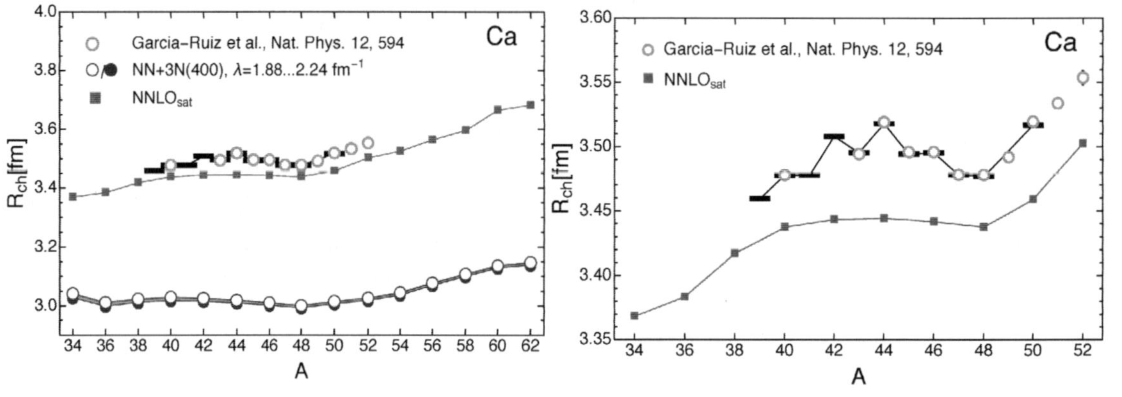

Figure 11. MR-IMSRG(2) charge radii of the calcium isotopes for NNLO_{sat} and NN+3N(400) with $\lambda = 1.88, \ldots, 2.24\,\text{fm}^{-1}$. Note the scale in the zoomed-in second panel. Black bars indicate experimental data [87, 121].

whose ground states we can converge in the MR-IMSRG(2) scheme for this specific input. This overbinding then goes along with an expected underestimation of nuclear charge radii, e.g., for the calcium isotopes shown in Fig. 11. NNLO_{sat}, on the other hand, is the first chiral NN+3N interaction that gives realistic binding energies and charge radii for ^{40}Ca and ^{48}Ca at the same time [123, 124]. Unfortunately, it is also a considerably harder interaction than NN+3N(400), and we already encounter convergence problems if we move a few isotopic chains past calcium.

Focusing on the calcium radii shown in the zoomed plot in the right panel of Fig. 11, we see that we do not reproduce the inverted parabolic trend between ^{40}Ca and ^{48}Ca. In traditional nuclear Shell Model/Configuration Interaction (CI) calculations with empirical interactions, this is explained due to configuration mixing of states containing 4p4h (Quadruples in chemistry parlance) and possibly higher excitations [125] that are not included in the MR-IMSRG(2) due to the truncation of the operators at the two-body level, and the use of spherical, particle-number projected HFB vacua as reference states in these calculations. We will discuss this issue in more depth in the next section.

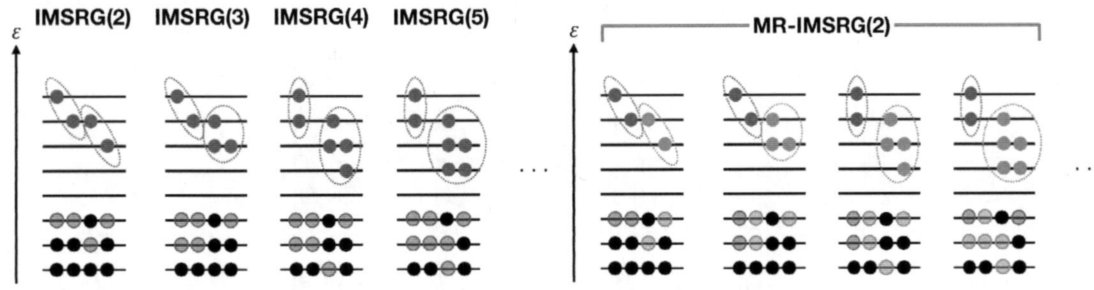

Figure 12. Schematic view of correlations in nuclei. Solid circles indicate nucleons, transparent circles hole states, and dashed ellipses indicate correlations between nucleons. The left panel indicates the IMSRG truncation that would be required to describe the sketched 2p2h (Doubles), 3p3h (Triples), etc. correlations. In the right panel, certain 2p2h as well as all 3p3h and higher correlations (indicated in blue) are built into a correlated wave function that then serves as the reference state for an MR-IMSRG(2) calculation.

6. Reference States with Static Correlation

As explained in Sec. 4, the IMSRG allows us to control which correlations are built into the reference state and the RG-improved Hamiltonian, respectively. This flexibility offers us an appealing option for addressing the lack of specific features in nuclear charge radii that we described at the end of Sec. 5.

Many-body bases built from a Slater determinant reference and its particle-hole excitations work best for systems with large gaps in the single-particle spectrum, such as closed-shell nuclei. If the gap is small, particle-hole excited basis states can be near-degenerate with the reference determinant, which results in strong configuration mixing. When the mixing involves configurations in which many nucleons are excited simultaneously, we will have *static* or *collective correlations* in the wave function, as opposed to *dynamic correlations* that result from a small number of nucleons only.

Important examples are the emergence of nuclear superfluidity [126] or diverse rotational and vibrational bands in open-shell nuclei (see, e.g., [127]). These phenomena are conveniently described by using the concept of intrinsic wave functions that explicitly break appropriate symmetries of the Hamiltonian. For instance, nuclear superfluidity can be treated to leading-order by introducing quasiparticles that are defined as superpositions of particle and holes [103, 128, 129]. The ground-state of an open-shell nucleus can be expressed as an antisymmetrized product state of these quasiparticles, which can then be optimized variationally — this is precisely the HFB method that was the first step in the generation of references states for the studies of isotopic chains discussed in Sec. 5. The intrinsic HFB wave functions are superpositions of states with different particle numbers, hence we eventually proceeded to restore the symmetry by particle number projection. Such approaches that involve symmetry breaking and subsequent restoration via projection have a long history in nuclear many-body theory [103, 130–140]).

The benefit of symmetry breaking and projection methods is that they capture static correlations in a set of simple one-body transition density matrices [103, 140]. Instead of using a Slater determinant reference state and trying to describe static correlations via 3p3h (Triples), 4p4h (Quadruples) and higher excitations of this reference via a high-rank IMSRG evolution (left panel of Fig. 12), we can use a correlated reference state whose irreducible density matrices $\lambda^{(k)}$ we can construct on the fly with little effort, and perform an MR-IMSRG(2) evolution to treat the dynamical correlation on top of this state (right panel of Fig. 12).

As a proof of principle, we show MR-IMSRG(2) calculations for the nucleus ^{20}Ne in Fig. 13.

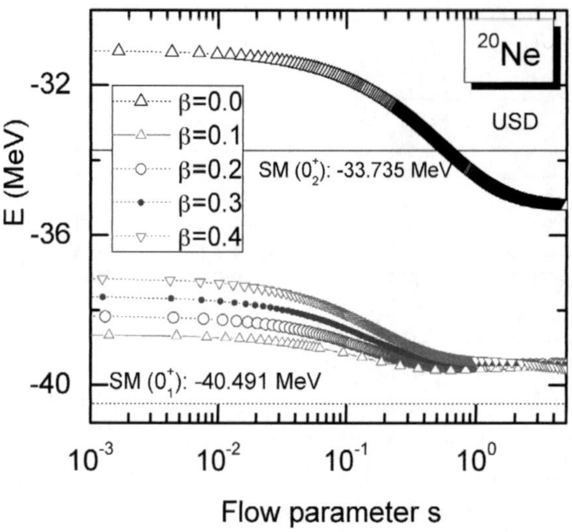

Figure 13. MR-IMSRG(2) energy of ^{20}Ne, starting from reference states with prolate intrinsic deformations $\beta \geq 0$ that are subsequently projected on angular momentum $J^{\pi} = 0^{+}$. These proof-of-principle calculations were performed with the USD Shell Model interaction [141, 142].

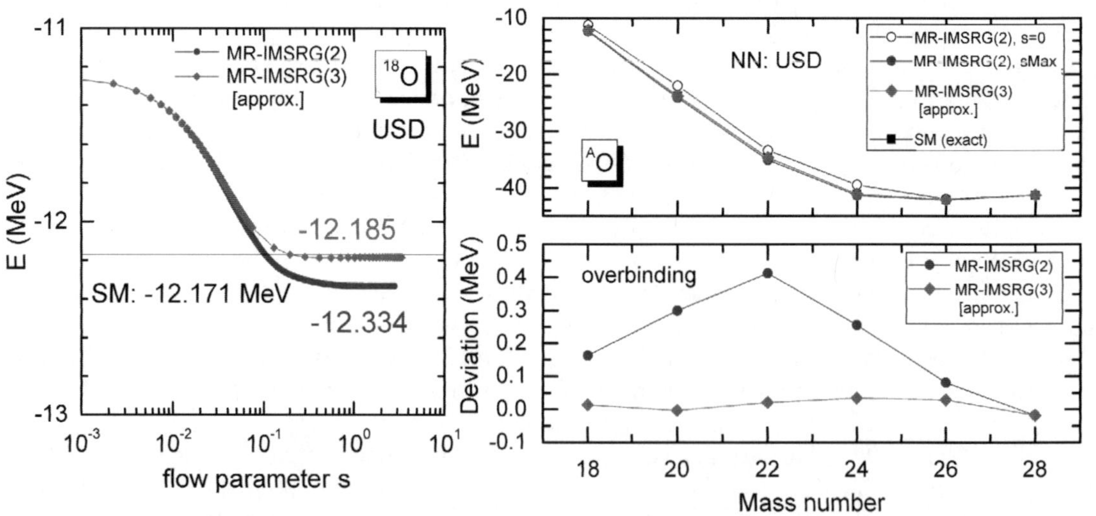

Figure 14. MR-IMSRG(2) and approximate MR-IMSRG(3) ground-state energies of the oxygen isotopes, using the USD Shell Model interaction [141, 142].

Both the proton and neutron shells are open in ^{20}Ne, which causes the ground-state to develop prolate intrinsic deformation. We obtain reference states by performing HFB calculations with a constraint on the quadrupole moment of the many-body state, and subsequently projecting them on good particle number and angular momentum $J^{\pi} = 0^{+}$. The references with prolate deformation ($\beta > 0.0$) flow towards the same 0^{+} ground state whose energy is close to the result of an exact diagonalization with the empirical USD Hamiltonian [141, 142] we use in our calculation. A spherical reference state ($\beta = 0.0$) flows to an excited 0^{+} state in the spectrum of ^{20}Ne instead.

The discrepancies between the MR-IMSRG(2) energies for the intrinsically prolate and

IOP Conf. Series: Journal of Physics: Conf. Series **1041** (2018) 012007 doi:10.1088/1742-6596/1041/1/012007

spherical states and the exact Shell Model results are caused by the truncation of the flow equations at the two-body level. While a full MR-IMSRG(3) calculation would have a computational cost of $\mathcal{O}(N^9)$, one-step approximations that leverage information from the MR-IMSRG(2) can be implemented at a more manageable $\mathcal{O}(N^7)$ effort [143], analogous to completely renormalized Coupled Cluster schemes that account for leading Triples excitations [144, 145]. In the left panel of Fig. 14, we perform such an approximate MR-IMSRG(3) calculation for ^{18}O, and find that we recover the exact diagonalization result[146]. In the right panel, we show that the same improvement is found for all even oxygen isotopes, which is very promising for future applications of the approximate MR-IMSRG(3) in general open-shell nuclei.

7. RG-Improved Hamiltonians: Merging (MR-)IMSRG with Other Many-Body Methods

In the previous sections we have focused on using the (MR-)IMSRG to directly calculate the ground-state energy of nuclei, expressed as the zero-body part of the RG-improved Hamiltonian in the limit $s \to \infty$. We are only realizing a fraction of the power and utility of the (MR-)IMSRG framework in this way: In principle, we have the entire Hamiltonian at our disposal, and not just at the end-point of the flow, but along an entire RG trajectory, which equips us with additional diagnostic power — we recall the variation of the triton ground-state energy with the resolution scale in a free-space SRG, and how we can draw conclusions about missing induced forces from it (see Sec. 3).

In our discussion of Fig. 6, we pointed out that the suppression of the off-diagonal matrix elements that couple the ground state to 2p2h excitations will also eliminate the outermost side diagonals of the Hamiltonian matrix, which implies that it can be beneficial to use the RG-improved Hamiltonian as an input for a secondary many-body calculation. We have explored this idea in several ways:

- With a simple modification of the definition of the off-diagonal Hamiltonian, it is possible to systematically derive effective interactions and operators for valence-space CI calculations, i.e., the traditional interacting Shell Model [95, 110, 111, 147]. This allows us to make use of existing technology to study any nucleus whose low-lying states can be converged in typical Shell Model calculations, explaining the coverage of the nuclear chart we have shown in Fig. 1. it also directly connects Shell Model phenomenology to the underlying chiral interactions — the valence-space diagonalization is merely an approximation to a no-core, large-scale diagonalization. In applications for *sd*-shell nuclei ($8 \leq Z \leq 20$), we were able to achieve a surprisingly good agreement with calculations using the gold standard empirical USD interaction [141, 142], which itself has an rms deviation of a mere $\sim 130\,\mathrm{keV}$ for more than 600 energy levels in this region of the nuclear chart.
- The (MR-)IMSRG was recently merged with the No-Core Shell Model (NCSM) into the so-called In-Medium NCSM, in an effort to combine the strengths of the two approaches [148]. A NCSM calculation in a small space provides a correlated reference state for the normal ordering and MR-IMSRG evolution, which enhances the Hamiltonian with dynamical correlation corresponding whose description would otherwise require extremely large NCSM model spaces. For NCSM reference states, the decoupling of the ground-state from excitations is not perfect, hence a secondary small NCSM diagonalization is performed to extract the low-lying eigenvalues and the associated eigenstates. In the IM-NCSM, we make explicit use of the Hamiltonian's flow-parameter dependence for diagnostic purposes (see Ref. [148] for more details).
- Finally, the ground-state decoupled Hamiltonian can be used as input for Equation-of-Motion (EOM) methods like the well-known Tamm-Dancoff (TDA) and Random Phase Approximations (RPA). TDA and RPA will actually become identical for an IMSRG

IOP Conf. Series: Journal of Physics: Conf. Series **1041** (2018) 012007 doi:10.1088/1742-6596/1041/1/012007

Hamiltonian, because the ground-state decoupling removes the feedback of RPA correlations into the ground state [149]. EOM calculations are complementary to the CI-based approaches discussed above because they allow us to explore the effects of large single-particle bases, at the cost of limiting the particle rank of the excitation operator for computational efficiency. Thus, they are best suited for the description of dynamical correlations in excited states, although the development of an MR-EOM approach based on correlated reference states is in progress.

The combination of the (MR-)IMSRG with CI/Shell Model techniques allows us to treat both dynamical and static correlation (see Sec. 6), but limits us through the factorial computational scaling of the diagonalization step. As already pointed out by White in Ref. [47], the Density Matrix Renormalization Group (DMRG) can be an appealing and efficient alternative to large-scale CI methods, especially when combined with (MR-)IMSRG decoupling (or Canonical Diagonalization in his choice of words). This idea was pursed by Yanai and co-workers in quantum chemistry [48, 98, 150–152]), and we are exploring its applications in nuclear physics.

Finally, we are looking ahead at applications of the IMSRG framework to the description of nuclei with cluster structures, and eventually reaction processes. Such structures and processes are most conveniently modeled on spatial lattices, which is why we are working on formulating the IMSRG in that context. Tensor RG methods that generalize the ideas of DMRG to higher dimensions have been very successful in describing correlated systems on spatial lattices could be an ideal complement to the IMSRG (see, e.g., [54–56]) . Interestingly, these Tensor RGs rely on unitary transformations to disentangle long- and short-range physics that are at least superficially reminiscent of SRG and IMSRG transformations — this certainly merits further investigation.

8. Conclusions and Outlook

In this contribution, we have attempted to illustrate the current reach and capabilities of modern *ab initio* nuclear structure theory, using the (MR-)IMSRG as a representative example. The discussion is by no means exhaustive, given the rapid rate at which new advances are being introduced. While our focus was on the many-body method, the presented results show that the input NN+3N interactions from chiral EFT are currently the dominant source of uncertainty in our theoretical results, but promising new chiral interactions [5, 8, 9, 17, 66, 153] are now being explored by the nuclear many-body community. The MR-IMSRG will be particularly useful in the community's efforts to confront these new Hamiltonians with current and forthcoming experimental data for medium-mass and heavy open-shell nuclei.

Work is currently underway to extend the MR-IMSRG framework in several directions, including the coupling to continuum degrees of freedom, and the treatment of intrinsic deformation. The present contribution showed first proof-of-principle results for the latter, implementing the MR-IMSRG transformation with an intrinsically deformed, angular-momentum projected reference state. We plan to extend this work further by using the Generator Coordinate Method to mix multiple projected configurations (see, e.g., [103]).

Motivated by our successes in using the (MR-)IMSRG to construct RG-improved Hamiltonians that can serve as input for the traditional nuclear Shell Model/CI [110, 111, 147], the No-Core Shell Model [148], and EOM methods [149], we seek to integrate the (MR-)IMSRG evolution with further approaches, in particular DMRG, which presents an appealing alternative to costly diagonalization methods for treating static correlations in configuration space. Such a combination of the IMSRG for the Hamiltonian with an entanglement-based RG for wave functions could also offer interesting opportunities for describing strongly interacting systems on spatial or spacetime lattices.

Acknowledgments
We are indebted to our collaborators on IMSRG projects, namely E. Gebrerufael, M. Hjorth-Jensen, J.D. Holt, R. Roth, A. Schwenk, S.R. Stroberg and K. Vobig, as well as K. Tsukiyama and F. Yuan. We also thank C. Barbieri, S. Binder, A. Calci, T. Duguet, F. Evangelista, R. J. Furnstahl, G. Hagen, K. Hebeler, G. R. Jansen, J. Simonis, V. Somà, and K. A. Wendt for many useful discussions, as well as providing interaction elements and/or theoretical results for comparison.

This publication is based on work supported in part by the National Science Foundation under Grants No. PHY-1404159, PHY-1614130, and PHY-1713901, as well as the U.S. Department of Energy, Office of Science, Office of Nuclear Physics under Grants No. DE-SC0008511 and DE-SC0008641 (NUCLEI SciDAC Collaboration), DE-FG02-97ER41019 and DE-SC0004142.

Computing resources were provided by the Michigan State University Institute for Cyber-Enabled Research (iCER), and the National Energy Research Scientific Computing Center (NERSC), a DOE Office of Science User Facility supported by the Office of Science of the U.S. Department of Energy under Contract No. DE-AC02-05CH11231.

References

[1] Detmold W 2015 *Nuclear Physics from Lattice QCD* (Springer International Publishing) pp 153–194 ISBN 978-3-319-08022-2 URL http://dx.doi.org/10.1007/978-3-319-08022-2_5

[2] Epelbaum E, Hammer H W and Meißner U G 2009 *Rev. Mod. Phys.* **81**(4) 1773–1825 URL http://link.aps.org/doi/10.1103/RevModPhys.81.1773

[3] Machleidt R and Entem D 2011 *Phys. Rept.* **503** 1 – 75 ISSN 0370-1573 URL http://www.sciencedirect.com/science/article/pii/S0370157311000457

[4] Epelbaum E, Krebs H and Meißner U G 2015 *Phys. Rev. Lett.* **115**(12) 122301 URL http://link.aps.org/doi/10.1103/PhysRevLett.115.122301

[5] Entem D R, Kaiser N, Machleidt R and Nosyk Y 2015 *Phys. Rev. C* **91**(1) 014002 URL http://link.aps.org/doi/10.1103/PhysRevC.91.014002

[6] Machleidt R and Sammarruca F 2016 *Physica Scripta* **91** 083007 URL http://stacks.iop.org/1402-4896/91/i=8/a=083007

[7] Gezerlis A, Tews I, Epelbaum E, Freunek M, Gandolfi S, Hebeler K, Nogga A and Schwenk A 2014 *Phys. Rev. C* **90**(5) 054323 URL http://link.aps.org/doi/10.1103/PhysRevC.90.054323

[8] Lynn J E, Tews I, Carlson J, Gandolfi S, Gezerlis A, Schmidt K E and Schwenk A 2016 *Phys. Rev. Lett.* **116**(6) 062501 URL http://link.aps.org/doi/10.1103/PhysRevLett.116.062501

[9] Lynn J E, Tews I, Carlson J, Gandolfi S, Gezerlis A, Schmidt K E and Schwenk A 2017 *Phys. Rev. C* **96**(5) 054007 URL https://link.aps.org/doi/10.1103/PhysRevC.96.054007

[10] Pastore S, Girlanda L, Schiavilla R, Viviani M and Wiringa R B 2009 *Phys. Rev. C* **80**(3) 034004 URL http://link.aps.org/doi/10.1103/PhysRevC.80.034004

[11] Pastore S, Girlanda L, Schiavilla R and Viviani M 2011 *Phys. Rev. C* **84**(2) 024001 URL http://link.aps.org/doi/10.1103/PhysRevC.84.024001

[12] Piarulli M, Girlanda L, Marcucci L E, Pastore S, Schiavilla R and Viviani M 2013 *Phys. Rev. C* **87**(1) 014006 URL http://link.aps.org/doi/10.1103/PhysRevC.87.014006

[13] Kölling S, Epelbaum E, Krebs H and Meißner U G 2009 *Phys. Rev. C* **80**(4) 045502 URL http://link.aps.org/doi/10.1103/PhysRevC.80.045502

[14] Kölling S, Epelbaum E, Krebs H and Meißner U G 2011 *Phys. Rev. C* **84**(5) 054008 URL http://link.aps.org/doi/10.1103/PhysRevC.84.054008

[15] Barnea N, Contessi L, Gazit D, Pederiva F and van Kolck U 2015 *Phys. Rev. Lett.* **114**(5) 052501 URL https://link.aps.org/doi/10.1103/PhysRevLett.114.052501

[16] Contessi L, Lovato A, Pederiva F, Roggero A, Kirscher J and van Kolck U 2017 *Physics Letters B* **772** 839–848 URL http://www.sciencedirect.com/science/article/pii/S0370269317306044

[17] Ekström A, Jansen G R, Wendt K A, Hagen G, Papenbrock T, Carlsson B D, Forssén C, Hjorth-Jensen M, Navrátil P and Nazarewicz W 2015 *Phys. Rev. C* **91**(5) 051301 URL http://link.aps.org/doi/10.1103/PhysRevC.91.051301

[18] Shirokov A M, Shin I J, Kim Y, Sosonkina M, Maris P and Vary J P 2016 *Physics Letters B* **761** 87–91 URL http://www.sciencedirect.com/science/article/pii/S0370269316304269

[19] Lepage G P 1989 (*Preprint* hep-ph/0506330)

[20] Lepage G P 1997 (*Preprint* nucl-th/9706029)

[21] Bogner S K, Kuo T T S and Schwenk A 2003 *Phys. Rept.* **386** 1–27 (*Preprint* nucl-th/0305035)

[22] Bogner S K, Furnstahl R J and Perry R J 2007 *Phys. Rev. C* **75** 061001(R) (*Preprint* nucl-th/0611045)

[23] Bogner S K, Furnstahl R J and Schwenk A 2010 *Prog. Part. Nucl. Phys.* **65** 94–147 (*Preprint* 0912.3688)

[24] Furnstahl R J and Hebeler K 2013 *Rept. Prog. Phys.* **76** 126301 URL http://stacks.iop.org/0034-4885/76/i=12/a=126301

[25] Glazek S D and Wilson K G 1993 *Phys. Rev. D* **48** 5863–5872

[26] Wegner F 1994 *Ann. Phys. (Leipzig)* **3** 77

[27] Bogner S K, Furnstahl R J, Ramanan S and Schwenk A 2006 *Nucl. Phys. A* **773** 203–220 URL http://www.sciencedirect.com/science/article/B6TVB-4K4WG34-2/2/218ed5e2e993065d0432f1ffeafdc50e

[28] Roth R and Langhammer J 2010 *Phys. Lett. B* **683** 272 – 277 ISSN 0370-2693 URL http://www.sciencedirect.com/science/article/pii/S037026930901507X

[29] Tichai A, Langhammer J, Binder S and Roth R 2016 *Physics Letters B* **756** 283–288 URL http://www.sciencedirect.com/science/article/pii/S0370269316002008

[30] Barrett B R, Navrátil P and Vary J P 2013 *Prog. Part. Nucl. Phys.* **69** 131 – 181 ISSN 0146-6410 URL http://www.sciencedirect.com/science/article/pii/S0146641012001184

[31] Jurgenson E D, Maris P, Furnstahl R J, Navrátil P, Ormand W E and Vary J P 2013 *Phys. Rev. C* **87**(5) 054312 URL http://link.aps.org/doi/10.1103/PhysRevC.87.054312

[32] Hergert H, Binder S, Calci A, Langhammer J and Roth R 2013 *Phys. Rev. Lett.* **110**(24) 242501 URL http://link.aps.org/doi/10.1103/PhysRevLett.110.242501

[33] Roth R, Calci A, Langhammer J and Binder S 2014 *Phys. Rev. C* **90**(2) 024325 URL http://link.aps.org/doi/10.1103/PhysRevC.90.024325

[34] Binder S, Langhammer J, Calci A and Roth R 2014 *Phys. Lett. B* **736** 119 – 123 ISSN 0370-2693 URL http://www.sciencedirect.com/science/article/pii/S0370269314004961

[35] Hagen G, Papenbrock T, Hjorth-Jensen M and Dean D J 2014 *Rept. Prog. Phys.* **77** 096302 URL http://stacks.iop.org/0034-4885/77/i=9/a=096302

[36] Hagen G, Hjorth-Jensen M, Jansen G R and Papenbrock T 2016 *Phys. Scripta* **91** 063006 URL `http://stacks.iop.org/1402-4896/91/i=6/a=063006`

[37] Hergert H 2017 *Phys. Scripta* **92** 023002 URL `http://stacks.iop.org/1402-4896/92/i=2/a=023002`

[38] Tsukiyama K, Bogner S K and Schwenk A 2011 *Phys. Rev. Lett.* **106** 222502

[39] Hergert H, Bogner S K, Binder S, Calci A, Langhammer J, Roth R and Schwenk A 2013 *Phys. Rev. C* **87**(3) 034307 URL `http://link.aps.org/doi/10.1103/PhysRevC.87.034307`

[40] Hergert H, Bogner S K, Morris T D, Schwenk A and Tsukiyama K 2016 *Physics Reports* **621** 165–222 URL `http://www.sciencedirect.com/science/article/pii/S0370157315005414`

[41] Kehrein S 2006 *The Flow Equation Approach to Many-Particle Systems* (*Springer Tracts in Modern Physics* vol 237) (Springer Berlin / Heidelberg)

[42] Heidbrink C and Uhrig G 2002 *Eur. Phys. J. B* **30**(4) 443–459 ISSN 1434-6028 10.1140/epjb/e2002-00401-9 URL `http://dx.doi.org/10.1140/epjb/e2002-00401-9`

[43] Drescher N A, Fischer T and Uhrig G S 2011 *Eur. Phys. J. B* **79**(2) 225–240 ISSN 1434-6028 10.1140/epjb/e2010-10723-6 URL `http://dx.doi.org/10.1140/epjb/e2010-10723-6`

[44] Krull H, Drescher N A and Uhrig G S 2012 *Phys. Rev. B* **86**(12) 125113 URL `http://link.aps.org/doi/10.1103/PhysRevB.86.125113`

[45] Fauseweh B and Uhrig G S 2013 *Phys. Rev. B* **87**(18) 184406 URL `http://link.aps.org/doi/10.1103/PhysRevB.87.184406`

[46] Krones J and Uhrig G S 2015 *Phys. Rev. B* **91**(12) 125102 URL `http://link.aps.org/doi/10.1103/PhysRevB.91.125102`

[47] White S R 2002 *J. Chem. Phys.* **117** 7472–7482 ISSN 00219606 URL `http://dx.doi.org/doi/10.1063/1.1508370`

[48] Yanai T and Chan G K L 2007 *J. Chem. Phys.* **127** 104107 (pages 14) URL `http://link.aip.org/link/?JCP/127/104107/1`

[49] Nakatsuji H 1976 *Phys. Rev. A* **14**(1) 41–50 URL `http://link.aps.org/doi/10.1103/PhysRevA.14.41`

[50] Mukherjee D and Kutzelnigg W 2001 *J. Chem. Phys.* **114** 2047–2061 URL `http://link.aip.org/link/?JCP/114/2047/1`

[51] Mazziotti D A 2006 *Phys. Rev. Lett.* **97**(14) 143002 URL `http://link.aps.org/doi/10.1103/PhysRevLett.97.143002`

[52] Evangelista F A 2014 *J. Chem. Phys.* **141** 054109 URL `http://scitation.aip.org/content/aip/journal/jcp/141/5/10.1063/1.4890660`

[53] White S R 1992 *Phys. Rev. Lett.* **69**(19) 2863–2866 URL `https://link.aps.org/doi/10.1103/PhysRevLett.69.2863`

[54] Vidal G 2007 *Phys. Rev. Lett.* **99**(22) 220405 URL `https://link.aps.org/doi/10.1103/PhysRevLett.99.220405`

[55] Evenbly G and Vidal G 2009 *Phys. Rev. Lett.* **102**(18) 180406 URL `https://link.aps.org/doi/10.1103/PhysRevLett.102.180406`

[56] Evenbly G and Vidal G 2015 *Phys. Rev. Lett.* **115**(18) 180405 URL `https://link.aps.org/doi/10.1103/PhysRevLett.115.180405`

[57] Gross D J and Wilczek F 1973 *Phys. Rev. Lett.* **30**(26) 1343–1346 URL `http://link.aps.org/doi/10.1103/PhysRevLett.30.1343`

[58] Politzer H D 1973 *Phys. Rev. Lett.* **30**(26) 1346–1349 URL http://link.aps.org/doi/10.1103/PhysRevLett.30.1346

[59] Weinberg S 1991 *Nuclear Physics B* **363** 3–18 URL http://www.sciencedirect.com/science/article/pii/055032139190231L

[60] Weinberg S 1996 *The Quantum Theory of Fields, Vol. II. Modern Applications* 2nd ed (UK: Cambridge University Press)

[61] Epelbaum E 2010 (*Preprint* 1001.3229)

[62] Heisenberg W 1932 *Zeitschrift für Physik* **77** 1–11 URL https://doi.org/10.1007/BF01342433

[63] Goldstone J 1961 *Il Nuovo Cimento (1955-1965)* **19** 154–164 URL https://doi.org/10.1007/BF02812722

[64] Melendez J A, Wesolowski S and Furnstahl R J 2017 *Phys. Rev. C* **96**(2) 024003 URL https://link.aps.org/doi/10.1103/PhysRevC.96.024003

[65] Gazit D, Quaglioni S and Navrátil P 2009 *Phys. Rev. Lett.* **103**(10) 102502 URL http://link.aps.org/doi/10.1103/PhysRevLett.103.102502

[66] Reinert P, Krebs H and Epelbaum E 2017 (*Preprint* 1711.08821)

[67] Valderrama M P, Sánchez M S, Yang C J, Long B, Carbonell J and van Kolck U 2017 *Phys. Rev. C* **95**(5) 054001 URL https://link.aps.org/doi/10.1103/PhysRevC.95.054001

[68] Sánchez M S, Yang C J, Long B and van Kolck U 2018 *Phys. Rev. C* **97**(2) 024001 URL https://link.aps.org/doi/10.1103/PhysRevC.97.024001

[69] Coester F, Cohen S, Day B and Vincent C M 1970 *Phys. Rev. C* **1**(3) 769–776 URL https://link.aps.org/doi/10.1103/PhysRevC.1.769

[70] Bethe H A 1971 *Ann. Rev. Nucl. Sci.* **21** 93–244 URL http://dx.doi.org/10.1146/annurev.ns.21.120171.000521

[71] Kuckei J, Montani F, Muther H and Sedrakian A 2003 *Nucl. Phys. A* **723** 32–44 (*Preprint* nucl-th/0210010)

[72] Hebeler K, Bogner S K, Furnstahl R J, Nogga A and Schwenk A 2011 *Phys. Rev. C* **83**(3) 031301 URL http://link.aps.org/doi/10.1103/PhysRevC.83.031301

[73] Hebeler K and Furnstahl R J 2013 *Phys. Rev. C* **87**(3) 031302 URL http://link.aps.org/doi/10.1103/PhysRevC.87.031302

[74] Wiringa R B, Stoks V G J and Schiavilla R 1995 *Phys. Rev. C* **51**(1) 38–51 URL https://link.aps.org/doi/10.1103/PhysRevC.51.38

[75] Carlson J, Gandolfi S, Pederiva F, Pieper S C, Schiavilla R, Schmidt K E and Wiringa R B 2015 *Rev. Mod. Phys.* **87**(3) 1067–1118 URL http://link.aps.org/doi/10.1103/RevModPhys.87.1067

[76] Entem D R and Machleidt R 2003 *Phys. Rev. C* **68** 041001 URL http://link.aps.org/doi/10.1103/PhysRevC.68.041001

[77] Roth R, Langhammer J, Calci A, Binder S and Navrátil P 2011 *Phys. Rev. Lett.* **107**(7) 072501 URL http://link.aps.org/doi/10.1103/PhysRevLett.107.072501

[78] Hergert H, Bogner S K, Morris T D, Binder S, Calci A, Langhammer J and Roth R 2014 *Phys. Rev. C* **90**(4) 041302 URL http://link.aps.org/doi/10.1103/PhysRevC.90.041302

[79] Hagen G, Papenbrock T, Dean D J and Hjorth-Jensen M 2010 *Phys. Rev. C* **82** 034330 URL http://link.aps.org/doi/10.1103/PhysRevC.82.034330

[80] Roth R, Binder S, Vobig K, Calci A, Langhammer J and Navrátil P 2012 *Phys. Rev. Lett.* **109**(5) 052501 URL http://link.aps.org/doi/10.1103/PhysRevLett.109.052501

[81] Binder S, Langhammer J, Calci A, Navrátil P and Roth R 2013 *Phys. Rev. C* **87**(2) 021303 URL http://link.aps.org/doi/10.1103/PhysRevC.87.021303

[82] Somà V, Duguet T and Barbieri C 2011 *Phys. Rev. C* **84**(6) 064317 URL http://link.aps.org/doi/10.1103/PhysRevC.84.064317

[83] Somà V, Barbieri C and Duguet T 2013 *Phys. Rev. C* **87**(1) 011303 URL http://link.aps.org/doi/10.1103/PhysRevC.87.011303

[84] Somà V, Barbieri C and Duguet T 2014 *Phys. Rev. C* **89**(2) 024323 URL http://link.aps.org/doi/10.1103/PhysRevC.89.024323

[85] Somà V, Cipollone A, Barbieri C, Navrátil P and Duguet T 2014 *Phys. Rev. C* **89**(6) 061301 URL http://link.aps.org/doi/10.1103/PhysRevC.89.061301

[86] Hebeler K 2012 *Phys. Rev. C* **85**(2) 021002 URL http://link.aps.org/doi/10.1103/PhysRevC.85.021002

[87] Wang M, Audi G, Wapstra A, Kondev F, MacCormick M, Xu X and Pfeiffer B 2012 *Chin. Phys. C* **36** 1603 URL http://stacks.iop.org/1674-1137/36/i=12/a=003

[88] Jurgenson E D, Navrátil P and Furnstahl R J 2009 *Phys. Rev. Lett.* **103** 082501

[89] Jurgenson E D, Navrátil P and Furnstahl R J 2011 *Phys. Rev. C* **83**(3) 034301 URL http://link.aps.org/doi/10.1103/PhysRevC.83.034301

[90] Wendt K A 2013 *Advances in the Application of the Similarity Renormalization Group to Strongly Interacting Systems* Ph.D. thesis The Ohio State University

[91] Griesshammer H W 2015 *in proceedings of the "8th International Workshop on Chiral Dynamics"* vol PoS(CD15) p 104 (*Preprint* 1511.00490)

[92] Kutzelnigg W and Mukherjee D 1997 *J. Chem. Phys.* **107** 432–449 URL http://link.aip.org/link/?JCP/107/432/1

[93] Kong L, Nooijen M and Mukherjee D 2010 *J. Chem. Phys.* **132** 234107 (pages 8) URL http://link.aip.org/link/?JCP/132/234107/1

[94] Hagen G, Papenbrock T, Dean D J, Schwenk A, Nogga A, Włoch M and Piecuch P 2007 *Phys. Rev. C* **76**(3) 034302 URL http://link.aps.org/doi/10.1103/PhysRevC.76.034302

[95] Tsukiyama K, Bogner S K and Schwenk A 2012 *Phys. Rev. C* **85**(6) 061304 URL http://link.aps.org/doi/10.1103/PhysRevC.85.061304

[96] Morris T D, Parzuchowski N M and Bogner S K 2015 *Phys. Rev. C* **92**(3) 034331 URL http://link.aps.org/doi/10.1103/PhysRevC.92.034331

[97] Shavitt I and Bartlett R J 2009 *Many-Body Methods in Chemistry and Physics: MBPT and Coupled-Cluster Theory* (Cambridge University Press)

[98] Yanai T and Chan G K L 2006 *J. Chem. Phys.* **124** 194106 (pages 16) URL http://link.aip.org/link/?JCP/124/194106/1

[99] Valdemoro C 1987 *Theory and Practice of the Spin-Adapted Reduced Hamiltonians (SRH)* (Dordrecht: Springer Netherlands) pp 275–288 ISBN 978-94-009-3855-7 URL http://dx.doi.org/10.1007/978-94-009-3855-7_14

[100] Kutzelnigg W and Mukherjee D 2002 *J. Chem. Phys.* **116** 4787–4801 URL http://link.aip.org/link/?JCP/116/4787/1

[101] Kutzelnigg W and Mukherjee D 2004 *J. Chem. Phys.* **120** 7340–7349 URL http://link.aip.org/link/?JCP/120/7340/1

[102] Kutzelnigg W and Mukherjee D 2004 *J. Chem. Phys.* **120** 7350–7368 URL http://link.aip.org/link/?JCP/120/7350/1

[103] Ring P and Schuck P 1980 *The Nuclear Many-Body Problem* 1st ed (Springer)

[104] Cipollone A, Barbieri C and Navrátil P 2013 *Phys. Rev. Lett.* **111**(6) 062501 URL http://link.aps.org/doi/10.1103/PhysRevLett.111.062501

[105] Cipollone A, Barbieri C and Navrátil P 2015 *Phys. Rev. C* **92**(1) 014306 URL http://link.aps.org/doi/10.1103/PhysRevC.92.014306

[106] Otsuka T, Suzuki T, Holt J D, Schwenk A and Akaishi Y 2010 *Phys. Rev. Lett.* **105**(3) 032501 URL http://link.aps.org/doi/10.1103/PhysRevLett.105.032501

[107] Hagen G, Hjorth-Jensen M, Jansen G R, Machleidt R and Papenbrock T 2012 *Phys. Rev. Lett.* **108**(24) 242501 URL http://link.aps.org/doi/10.1103/PhysRevLett.108.242501

[108] Epelbaum E, Krebs H, Lähde T A, Lee D, Meißner U G and Rupak G 2014 *Phys. Rev. Lett.* **112**(10) 102501 URL http://link.aps.org/doi/10.1103/PhysRevLett.112.102501

[109] Holt J, Menéndez J and Schwenk A 2013 *Eur. Phys. J. A* **49** 1–6 ISSN 1434-6001 URL http://dx.doi.org/10.1140/epja/i2013-13039-2

[110] Bogner S K, Hergert H, Holt J D, Schwenk A, Binder S, Calci A, Langhammer J and Roth R 2014 *Phys. Rev. Lett.* **113**(14) 142501 URL http://link.aps.org/doi/10.1103/PhysRevLett.113.142501

[111] Stroberg S R, Hergert H, Holt J D, Bogner S K and Schwenk A 2016 *Phys. Rev. C* **93**(5) 051301 URL http://link.aps.org/doi/10.1103/PhysRevC.93.051301

[112] Hergert H and Roth R 2009 *Phys. Rev. C* **80**(2) 024312 URL http://link.aps.org/doi/10.1103/PhysRevC.80.024312

[113] Roth R 2009 *Phys. Rev. C* **79** 064324 URL http://link.aps.org/doi/10.1103/PhysRevC.79.064324

[114] Taube A G and Bartlett R J 2008 *J. Chem. Phys.* **128** 044110 (pages 13) URL http://link.aip.org/link/?JCP/128/044110/1

[115] Taube A G and Bartlett R J 2008 *J. Chem. Phys.* **128** 044111 (pages 9) URL http://link.aip.org/link/?JCP/128/044111/1

[116] Lee D 2009 *Prog. Part. Nucl. Phys.* **63** 117 – 154 ISSN 0146-6410 URL http://www.sciencedirect.com/science/article/pii/S014664100800094X

[117] Hoffman C R, Baumann T, Bazin D, Brown J, Christian G, DeYoung P A, Finck J E, Frank N, Hinnefeld J, Howes R, Mears P, Mosby E, Mosby S, Reith J, Rizzo B, Rogers W F, Peaslee G, Peters W A, Schiller A, Scott M J, Tabor S L, Thoennessen M, Voss P J and Williams T 2008 *Phys. Rev. Lett.* **100**(15) 152502 URL http://link.aps.org/doi/10.1103/PhysRevLett.100.152502

[118] Caesar C, Simonis J, Adachi T, Aksyutina Y, Alcantara J, Altstadt S, Alvarez-Pol H, Ashwood N, Aumann T, Avdeichikov V, Barr M, Beceiro S, Bemmerer D, Benlliure J, Bertulani C A, Boretzky K, Borge M J G, Burgunder G, Caamano M, Casarejos E, Catford W, Cederkäll J, Chakraborty S, Chartier M, Chulkov L, Cortina-Gil D, Datta Pramanik U, Diaz Fernandez P, Dillmann I, Elekes Z, Enders J, Ershova O, Estrade A, Farinon F, Fraile L M, Freer M, Freudenberger M, Fynbo H O U, Galaviz D, Geissel H, Gernhäuser R, Golubev P, Gonzalez Diaz D, Hagdahl J, Heftrich T, Heil M, Heine M, Heinz A, Henriques A, Holl M, Holt J D, Ickert G, Ignatov A, Jakobsson B, Johansson H T, Jonson B, Kalantar-Nayestanaki N, Kanungo R, Kelic-Heil A, Knöbel R, Kröll T, Krücken R, Kurcewicz J, Labiche M, Langer C, Le Bleis T, Lemmon R, Lepyoshkina O, Lindberg S, Machado J, Marganiec J, Maroussov V, Menéndez J, Mostazo M, Movsesyan A, Najafi A, Nilsson T, Nociforo C, Panin V, Perea A, Pietri S, Plag R, Prochazka A, Rahaman A, Rastrepina G, Reifarth R, Ribeiro G, Ricciardi M V, Rigollet C, Riisager K, Röder

M, Rossi D, Sanchez del Rio J, Savran D, Scheit H, Schwenk A, Simon H, Sorlin O, Stoica V, Streicher B, Taylor J, Tengblad O, Terashima S, Thies R, Togano Y, Uberseder E, Van de Walle J, Velho P, Volkov V, Wagner A, Wamers F, Weick H, Weigand M, Wheldon C, Wilson G, Wimmer C, Winfield J S, Woods P, Yakorev D, Zhukov M V, Zilges A, Zoric M and Zuber K (R3B collaboration) 2013 *Phys. Rev. C* **88**(3) 034313 URL http://link.aps.org/doi/10.1103/PhysRevC.88.034313

[119] Rogers W F, Garrett S, Grovom A, Anthony R E, Aulie A, Barker A, Baumann T, Brett J J, Brown J, Christian G, DeYoung P A, Finck J E, Frank N, Hamann A, Haring-Kaye R A, Hinnefeld J, Howe A R, Islam N T, Jones M D, Kuchera A N, Kwiatkowski J, Lunderberg E M, Luther B, Meyer D A, Mosby S, Palmisano A, Parkhurst R, Peters A, Smith J, Snyder J, Spyrou A, Stephenson S L, Strongman M, Sutherland B, Taylor N E and Thoennessen M 2015 *Phys. Rev. C* **92**(3) 034316 URL http://link.aps.org/doi/10.1103/PhysRevC.92.034316

[120] Kohley Z, Baumann T, Christian G, DeYoung P A, Finck J E, Frank N, Luther B, Lunderberg E, Jones M, Mosby S, Smith J K, Spyrou A and Thoennessen M 2015 *Phys. Rev. C* **91**(3) 034323 URL http://link.aps.org/doi/10.1103/PhysRevC.91.034323

[121] Angeli I and Marinova K P 2013 *Atomic Data and Nuclear Data Tables* **99** 69–95 URL http://www.sciencedirect.com/science/article/pii/S0092640X12000265

[122] Lapoux V, Somà V, Barbieri C, Hergert H, Holt J D and Stroberg S R 2016 *Phys. Rev. Lett.* **117**(5) 052501 URL http://link.aps.org/doi/10.1103/PhysRevLett.117.052501

[123] Hagen G, Ekström A, Forssén C, Jansen G R, Nazarewicz W, Papenbrock T, Wendt K A, Bacca S, Barnea N, Carlsson B, Drischler C, Hebeler K, Hjorth-Jensen M, Miorelli M, Orlandini G, Schwenk A and Simonis J 2015 *Nat. Phys.* **12** 186 URL http://dx.doi.org/10.1038/nphys3529

[124] Garcia Ruiz R F, Bissell M L, Blaum K, Ekstrom A, Frommgen N, Hagen G, Hammen M, Hebeler K, Holt J D, Jansen G R, Kowalska M, Kreim K, Nazarewicz W, Neugart R, Neyens G, Nortershauser W, Papenbrock T, Papuga J, Schwenk A, Simonis J, Wendt K A and Yordanov D T 2016 *Nat. Phys.* **12** 594–598 URL http://dx.doi.org/10.1038/nphys3645

[125] Caurier E, Langanke K, Martínez-Pinedo G, Nowacki F and Vogel P 2001 *Physics Letters B* **522** 240–244 URL http://www.sciencedirect.com/science/article/pii/S0370269301012461

[126] Dean D J and Hjorth-Jensen M 2003 *Rev. Mod. Phys.* **75** 607–656 (*Preprint* nucl-th/0210033)

[127] Bohr A and Mottelson B R 1999 *Nuclear Structure, Vol. II: Nuclear Deformations* (World Scientific)

[128] Bardeen J, Cooper L N and Schrieffer J R 1957 *Phys. Rev.* **106** 162

[129] Bogoliubov N N 1958 *Sov. Phys. JETP* **7** 41

[130] Peierls R E 1973 *Proc. R. Soc. London A* **333** 157–170 URL http://rspa.royalsocietypublishing.org/content/333/1593/157

[131] Egido J L and Ring P 1982 *Nucl. Phys. A* **383** 189–204

[132] Robledo L M 1994 *Phys. Rev. C* **50**(6) 2874–2881 URL http://link.aps.org/doi/10.1103/PhysRevC.50.2874

[133] Flocard H and Onishi N 1997 *Ann. Phys.* **254** 275

[134] Sheikh J A and Ring P 2000 *Nucl. Phys. A* **665** 71–91 (*Preprint* nucl-th/9907065)

[135] Dobaczewski J, Stoitsov M V, Nazarewicz W and Reinhard P G 2007 *Phys. Rev. C* **76** 054315 (*Preprint* 0708.0441)

IOP Conf. Series: Journal of Physics: Conf. Series **1041** (2018) 012007 doi:10.1088/1742-6596/1041/1/012007

[136] Bender M, Duguet T and Lacroix D 2009 *Phys. Rev. C* **79** 044319 (*Preprint* 0809.2045)

[137] Duguet T, Bender M, Bennaceur K, Lacroix D and Lesinski T 2009 *Phys. Rev. C* **79** 044320 (*Preprint* 0809.2049)

[138] Lacroix D, Duguet T and Bender M 2009 *Phys. Rev. C* **79** 044318 (*Preprint* 0809.2041)

[139] Lacroix D and Gambacurta D 2012 *Phys. Rev. C* **86**(1) 014306 URL http://link.aps.org/doi/10.1103/PhysRevC.86.014306

[140] Duguet T 2015 *J. Phys. G* **42** 025107 URL http://stacks.iop.org/0954-3899/42/i=2/a=025107

[141] Wildenthal B H 1984 *Progress in Particle and Nuclear Physics* **11** 5–51 URL http://www.sciencedirect.com/science/article/pii/0146641084900115

[142] Brown B A and Wildenthal B H 1988 *Annual Review of Nuclear and Particle Science* **38** 29–66 URL https://doi.org/10.1146/annurev.ns.38.120188.000333

[143] Morris T D 2016 *Systematic Improvements of Ab Initio In-Medium Similarity Renormalization Group Calculations* Ph.D. thesis Michigan State University URL http://etd.lib.msu.edu/islandora/object/etd%3A3864

[144] Piecuch P and Włoch M 2005 *J. Chem. Phys.* **123** 224105 URL http://scitation.aip.org/content/aip/journal/jcp/123/22/10.1063/1.2137318

[145] Binder S, Piecuch P, Calci A, Langhammer J, Navrátil P and Roth R 2013 *Phys. Rev. C* **88**(5) 054319 URL http://link.aps.org/doi/10.1103/PhysRevC.88.054319

[146] Brown B A and Richter W A 2006 *Phys. Rev. C* **74**(3) 034315 URL http://link.aps.org/doi/10.1103/PhysRevC.74.034315

[147] Stroberg S R, Calci A, Hergert H, Holt J D, Bogner S K, Roth R and Schwenk A 2017 *Phys. Rev. Lett.* **118**(3) 032502 URL http://link.aps.org/doi/10.1103/PhysRevLett.118.032502

[148] Gebrerufael E, Vobig K, Hergert H and Roth R 2017 *Phys. Rev. Lett.* **118**(15) 152503 URL https://link.aps.org/doi/10.1103/PhysRevLett.118.152503

[149] Parzuchowski N M, Morris T D and Bogner S K 2017 *Phys. Rev. C* **95**(4) 044304 URL https://link.aps.org/doi/10.1103/PhysRevC.95.044304

[150] Neuscamman E, Yanai T and Chan G K L 2009 *J. Chem. Phys.* **130** 124102 (pages 12) URL http://link.aip.org/link/?JCP/130/124102/1

[151] Neuscamman E, Yanai T and Chan G K L 2010 *J. Chem. Phys.* **132** 024106 (pages 13) URL http://link.aip.org/link/?JCP/132/024106/1

[152] Yanai T, Kurashige Y, Neuscamman E and Chan G K L 2010 *J. Chem. Phys.* **132** 024105 (pages 9) URL http://link.aip.org/link/?JCP/132/024105/1

[153] Entem D R, Kaiser N, Machleidt R and Nosyk Y 2015 *Phys. Rev. C* **92**(6) 064001 URL http://link.aps.org/doi/10.1103/PhysRevC.92.064001

IOP Conf. Series: Journal of Physics: Conf. Series **1041** (2018) 012008 doi:10.1088/1742-6596/1041/1/012008

Toward electrodynamics of unconventional phases of dilute nuclear matter

Armen Sedrakian

Frankfurt Institute for Advanced Studies, D-60438 Frankfurt am Main, Germany

E-mail: sedrakian@fias.uni-frankfurt.de

John W. Clark

Department of Physics and McDonnell Center for the Space Sciences, Washington University, St. Louis, Missouri 63130, USA

Centro de Investigação em Matemática e Aplicações, University of Madeira, 9020-105 Funchal, Madeira, Portugal

E-mail: jwc@wuphys.wustl.edu

Abstract. The phase diagram of isospin-asymmetrical nuclear matter may feature a number of unconventional phases, which include the translationally and rotationally symmetric, but isospin-asymmetrical BCS condensate, the current-carrying Larkin-Ovchinnikov-Fulde-Ferrell (LOFF) phase, and the heterogeneous phase-separated phase. Because the Cooper pairs of the condensate carry a single unit of charge, these phases are charged superconductors and respond to electromagnetic gauge fields by either forming domains (type-I superconductivity) or quantum vortices (type-II superconductivity). We evaluate the Ginzburg-Landau (GL) parameter across the phase diagram and find that the unconventional phases of isospin-asymmetrical nuclear matter are good type-II superconductors and should form Abrikosov vortices with twice the quantum of magnetic flux. We also find that the LOFF phase at the boundary of the transition to the type-I state, with the GL parameter being close to the critical value $1/\sqrt{2}$.

1. Introduction

Understanding superconducting phases of nuclear matter at sub-saturation densities is a challenging many-body physics problem. The attraction in the nuclear force is responsible for the formation of nuclear clusters, as well as condensates of Bardeen-Cooper-Schrieffer (BCS) type at low temperatures, with significant implications for the physics of supernovae and neutron stars. It has been conjectured such systems can exhibit the celebrated BCS-BEC transition [1], i.e., a transformation of loosely-bound Cooper pairs at weak coupling to bound dimers at strong coupling that a form a Bose-Einstein condensate (BEC) [2, 3, 4, 5, 6, 7, 8, 9, 10]. However, these nuclear systems are isospin asymmetric, which means that the isoscalar neutron-proton (np) pairing is disrupted by the mismatch in the Fermi surfaces of protons and neutrons [11].

In a recent series of papers [12, 13, 14], the combined effects of the BCS-BEC crossover and mismatched Fermi surfaces have been studied in the phase diagram of nuclear and neutron matter, see also the reviews [15, 16]. The coupled equations for the gap and the densities of the constituents (neutrons and protons) have been solved allowing for (i) the ordinary BCS state,

(ii) its low-density asymptotic counterpart BEC state, and (iii) two phases that emerge once there is finite isospin asymmetry, namely the current-carrying phase (LOFF phase) and the spatially separated (PS-BCS) phase. The strong-coupling regime features the phase-separated BEC (PS-BEC) phase. Another important characteristic of these phases – the coherence length ξ – was extracted microscopically from the wave functions of the Cooper pairs, thus avoiding the standard definition which is valid only in the BCS limit.

The present contribution focuses on the electromagnetic response of the phases indicated above, which arises from the fact that the proton in an np Cooper pair carries a unit of charge. For this purpose we evaluate the penetration depth λ of the magnetic field in various regions of the phase diagram and deduce the Ginzburg-Landau (GL) parameter $\kappa = \lambda/\xi$. We then use the GL criterion to establish the type of the superconductivity realized in the phase diagram.

We note here that the type-II versus type-I superconductivity of the S-wave proton condensate in the core of a neutron star has been widely discussed in the literature, see for the recent work [17, 18, 19, 20, 21, 22] and for a recent review Ref. [23]. The proton superfluid contributes at most 10-15 percent to the density of the core, therefore it actual density is of the same order as that of the neutron-proton condensate considered here. The main differences arise through larger gaps in the present case, which lead to smaller coherence lengths, and the charges that the condensate carry (a unit of charge in the present case).

This contribution is structured as follows. In Sec. 2 we give a brief review of the theory of asymmetrical nuclear matter and discuss the resulting phase diagram. In Sec. 3 we address the electrodynamics of the superconducting state; we evaluate the GL parameter using input from our previous microscopic studies along with a new computation of the penetration depth. In Section 4 we collect our conclusions and consider related perspectives.

2. Review of the phase diagram

Superfluid nuclear matter can be described in terms of the Nambu-Gor'kov propagators, which from a matrix

$$i\begin{pmatrix} G^{+}_{12} & F^{-}_{12} \\ F^{+}_{12} & G^{-}_{12} \end{pmatrix} = \begin{pmatrix} \langle T_\tau \psi_1 \psi_2^{+} \rangle & \langle T_\tau \psi_1 \psi_2 \rangle \\ \langle T_\tau \psi_1^{+} \psi_2^{+} \rangle & \langle T_\tau \psi_1^{+} \psi_2 \rangle \end{pmatrix}, \tag{1}$$

where $x = (t, \boldsymbol{r})$ denotes the continuous temporal-spatial variable and Greek indices label discrete spin and isospin variables. Each operator in Eq. (1) can be viewed as a bi-spinor, i.e., $\psi_\alpha = (\psi_{n\uparrow}, \psi_{n\downarrow}, \psi_{p\uparrow}, \psi_{p\downarrow})^T$, where the symbols $\uparrow, \downarrow$ label a particle's spin and the indices n, p label its isospin.

In energy-momentum space the elements of the matrix (1) are given by

$$G^{\pm}_{n/p} = \frac{ik_\nu \pm \epsilon^{\mp}_{p/n}}{(ik_\nu - E^{+}_{\mp/\pm})(ik_\nu + E^{-}_{\pm/\mp})}, \tag{2}$$

$$F^{\pm}_{np} = \frac{-i\Delta}{(ik_\nu - E^{+}_{\pm})(ik_\nu + E^{-}_{\mp})}, \tag{3}$$

$$F^{\pm}_{pn} = \frac{i\Delta}{(ik_\nu - E^{+}_{\mp})(ik_\nu + E^{-}_{\pm})}, \tag{4}$$

where the four branches of the quasiparticle spectrum are given by

$$E^{a}_{r} = \sqrt{E_S^2 + \Delta^2} + r\delta\mu + aE_A, \tag{5}$$

in which $a, r \in \{+, -\}$. Here

$$E_S = \frac{Q^2/4 + k^2}{2m^*} - \bar{\mu}, \quad E_A = \frac{\boldsymbol{k} \cdot \boldsymbol{Q}}{2m^*}, \tag{6}$$

are the symmetrical and anti-symmetrical parts of the quasiparticle spectrum, $\bar{\mu} \equiv (\mu_n + \mu_p)/2$, m^* is the effective mass of nucleon, and $\boldsymbol{Q}$ is the center-of-mass momentum of the Cooper pairs.

The anomalous self-energy (pairing-gap) evaluated in the standard fashion to lowest order in the neutron-proton interaction $V(\boldsymbol{k}, \boldsymbol{k}')$ is expressed as

$$\begin{aligned}\Delta(\boldsymbol{k}, \boldsymbol{Q}) &= \frac{1}{4\beta} \int \frac{d^3k'}{(2\pi)^3} \sum_{\nu} V(\boldsymbol{k}, \boldsymbol{k}') \\ &\quad \mathrm{Im}\Big[F^{+}_{np}(k'_\nu, \boldsymbol{k}', \boldsymbol{Q}) + F^{-}_{np}(k'_\nu, \boldsymbol{k}', \boldsymbol{Q}) - F^{+}_{pn}(k'_\nu, \boldsymbol{k}', \boldsymbol{Q}) - F^{+}_{pn}(k'_\nu, \boldsymbol{k}', \boldsymbol{Q})\Big], \end{aligned} \tag{7}$$

which yields, after partial-wave expansion and computation of the Matsubara sums,

$$\Delta_l(Q) = \frac{1}{4} \sum_{a,r,l'} \int \frac{d^3k'}{(2\pi)^3} V_{l,l'}(k, k') \frac{\Delta_{l'}(k', Q)}{2\sqrt{E_S^2(k') + \Delta^2(k', Q)}} [1 - 2f(E_a^r)], \tag{8}$$

where $V_{l,l'}(k, k')$ is now the interaction in the ${}^3S_1$-${}^3D_1$ partial wave, $f(\omega) = 1/[\exp(\omega/T) + 1]$, and $\Delta^2 = \sum_l \Delta_l^2$.

The densities of neutrons and protons are obtained from the propagator after phase-space integration and summation over the discrete (spin-isospin) variables:

$$\begin{aligned}\rho_{n/p}(\boldsymbol{Q}) &= \frac{2}{\beta} \int \frac{d^3k}{(2\pi)^3} \sum_{\nu} G^{+}_{n/p}(k_\nu, \boldsymbol{k}, \boldsymbol{Q}) \\ &= 2 \int \frac{d^3k}{(2\pi)^3} \left[\frac{1}{2} \left(1 + \frac{E_S}{\sqrt{E_S^2 + \Delta^2}} \right) f(E^{+}_{\mp}) + \frac{1}{2} \left(1 - \frac{E_S}{\sqrt{E_S^2 + \Delta^2}} \right) f(-E^{-}_{\pm}) \right]. \end{aligned} \tag{9}$$

In writing the thermodynamic quantities we distinguish the spatially homogeneous and inhomogeneous cases and label the superfluid (S) or unpaired (N) quantities correspondingly. The free energies of the homogeneous phases can then be simply written as

$$F_S = E_S - TS_S, \quad F_N = E_N - TS_N, \tag{10}$$

where E is the internal energy (statistical average of the system Hamiltonian) and S denotes the entropy. The free energy of the heterogeneous superfluid phase (i.e. the phase in which there is a separation of the normal and superfluid phases) is given by

$$\mathscr{F}(x, \alpha) = (1 - x)F_S(\alpha = 0) + xF_N(\alpha \neq 0), \ (Q = 0), \tag{11}$$

where x here denotes the filling fraction of the unpaired component and

$$\alpha = \frac{\rho_n - \rho_p}{\rho_n + \rho_p} \tag{12}$$

is the density asymmetry. The net density is given by $\rho = \rho_n + \rho_p$.

There are four possible states that can appear in the temperature-density phase diagram:

$$\begin{array}{llll} Q = 0, & \Delta \neq 0, & x = 0, & \text{BCS phase,} \\ Q \neq 0, & \Delta \neq 0, & x = 0, & \text{LOFF phase,} \\ Q = 0, & \Delta \neq 0, & x \neq 0, & \text{PS phase,} \\ Q = 0, & \Delta = 0, & x = 1, & \text{unpaired phase.} \end{array} \tag{13}$$

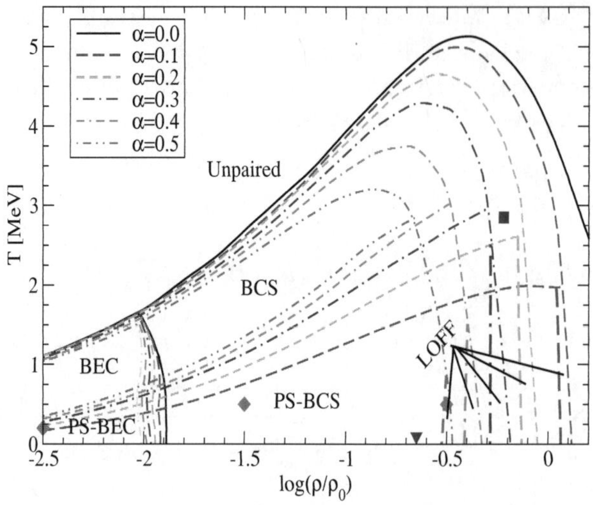

Figure 1. Phase diagram of dilute nuclear matter in the temperature-density plane for several isospin asymmetries α according to Refs. [12, 13]. The density is normalized to the nuclear saturation density. The phase diagram includes the unpaired phase, the BCS (BEC) phase, the LOFF phase, and the PS-BCS (PS-BEC) phase. For each asymmetry there are two tri-critical points, one of which is always a Lifshitz point [12, 13]. There is a single tetra-critical point in the phase diagram shown by the square dot. Note that the LOFF phase exists in the lower right corner of the phase diagram. The boundaries between BCS and BEC phases are identified by the change of sign of the average chemical potential $\bar{\mu}$. The diamonds mark the values of the density and temperature appearing in Table 1, these being chosen as representative of the three coupling regimes.

Clearly, the phase with lowest free energy at any given temperature and density corresponds to the ground state of the nuclear matter. Their locations in the phase diagram are shown in Fig. 1 for several values of isospin asymmetry α. One can identify four different phases classified according to Eq. (13): (a) The unpaired phase occupies the high-temperature region $T > T_{c0}$ of the phase diagram, where $T_{c0}(\rho)$ is the critical temperature of the normal/superfluid phase transition at $\alpha = 0$. (b) The current-carrying LOFF phase is located in the lower right corner, corresponding to low temperatures and high densities. (c) The PS phase occupies the lower left corner, which corresponds to low densities and low temperatures. As the temperature is increased, these last two phases transform into (d) the isospin-asymmetrical BCS phase.

To quantify the BCS-BEC crossover, we define three regimes of coupling which we express in terms of densities. The strong-coupling regime (SCR) corresponds to the low-density limit where well-defined deuterons exist. The weak-coupling regime (WCR) corresponds to the high-density limit where well-defined Cooper pairs are formed. The regime intermediate between these domains is called the intermediate-coupling regime (ICR). Representative points in these regimes are indicated by diamonds in Fig. 1.

3. Evaluating the Ginzburg-Landau parameter

BCS superconductors are characterized by at least three distinct length scales: (i) the *London penetration depth* λ, (ii) the *coherence length* ξ, and (iii) the mean interparticle distance d. The ratio of two of these scales defines the Ginzburg-Landau (GL) parameter: $\kappa = \lambda/\xi$. In the range $1/\sqrt{2} < \kappa < \infty$, the material is a type-II superconductor; otherwise it is type-I. In conventional

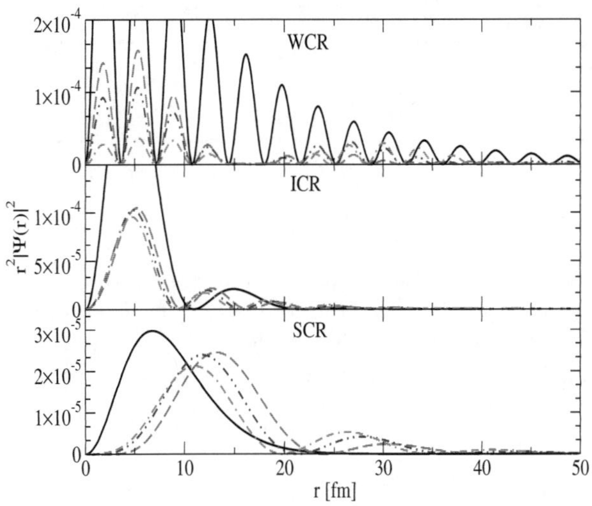

Figure 2. Dependence of $r^2|\Psi(r)|^2$ on r for the three coupling regimes [13]. (See Table 1 for corresponding values of density and temperature). For the asymmetry parameter we consider values $\alpha = 0$ (black, solid), 0.1 (red, long-dashed), 0.2 (blue, dash–double-dotted) and 0.3 (violet, double-dash-dotted).

type-II superconductors with Cooper charge $2e$ the magnetic field is carried by electromagnetic vortices with quantum flux $\phi_0 = \pi\hbar c/e$.

We now review the evaluation the coherence length, which can be related to the root-mean-square radius of the Cooper-pair wave function, defined as [12, 13]

$$\langle r^2 \rangle = \int d^3r \, r^2 |\Psi(\boldsymbol{r})|^2, \tag{14}$$

where $\Psi(\boldsymbol{r})$ is the Cooper-pair wave function. Fig. 2 shows the integrand of Eq. (14), i.e., the quantity $r^2|\Psi(\boldsymbol{r})|^2$. The spatial correlation in the SCR is found to display a single peak for $\alpha = 0$, which clearly indicates a tightly bound state at the origin. The residual oscillations (for finite asymmetries) indicate that there is no single bound state. The increasingly oscillatory behavior of the function $r^2|\Psi(\boldsymbol{r})|^2$ in the ICR is consistent with the notion of a transition from the BEC to the BCS regime. In the WCR, strong oscillations are observed, reflecting the coherence of the BCS state over large distances.

The coherence length, i.e., the spatial extension of a Cooper pair, is then defined as

$$\xi_{\rm rms} = \sqrt{\langle r^2 \rangle}. \tag{15}$$

In the weak-coupling BCS regime the coherence length is given by the well-known analytical formula

$$\xi_a = \frac{\hbar^2 k_F}{\pi m^* \Delta}. \tag{16}$$

Table 1 lists the values of coherence length in the three coupling regimes, as obtained from Eq. (15) and through the BCS formula (16). The corresponding values of the mean interparticle distance listed in the Table are determined from the density.

We now apply the macroscopic equations of electrodynamics of superconductors. Since neutron-proton Cooper pairs carry a single unit of charge rather than two, the relations between

	$\log_{10}(\rho/\rho_0)$	k_F[fm^{-1}]	T [MeV]	d [fm]	$\xi_{\rm rms}$ [fm]	ξ_a [fm]	λ [fm]	$\kappa_{\rm rms}$	$\kappa_{\rm rms}^{\rm LOFF}$
WCR	−0.5	0.91	0.5	1.68	3.17	1.41	18.5	5.8	0.8
ICR	−1.5	0.42	0.5	3.61	0.94	1.25	30.5	32.4	–
SCR	−2.5	0.20	0.2	7.79	0.57	1.79	50.5	88.2	–

Table 1. For each of the three regimes of coupling strength, corresponding values are presented for the density ρ (in units of nuclear saturation density $\rho_0 = 0.16$ fm^{-3}), Fermi momentum k_F, temperature T, mean interparticle distance d, and coherence parameters $\xi_{\rm rms}$ and ξ_a. At $\alpha = 0$ the values of the pairing gap Δ in the three regimes WCR, ICR, and SCR are respectively 9.39, 4.50, and 1.44 MeV, while the corresponding values of the effective mass (in units of the bare mass) are 0.903, 0.989, and 0.999. In the regime WCR, the LOFF phase is found in the vicinity of asymmetry $\alpha = 0.49$, at which $\Delta = 1.27$ MeV and $Q = 0.4$ fm^{-3}.

various quantities differ accordingly from those for ordinary superconductors. To keep track of these differences we denote the charge and mass of a Cooper pair by e_X and m_X.

The gauge invariant momentum associated with the Cooper wave function is given by

$$\boldsymbol{p} = m_X \boldsymbol{v} = \hbar \boldsymbol{\nabla} \chi - \frac{e_X}{c} \boldsymbol{A}, \tag{17}$$

where χ the phase of the wave function and $\boldsymbol{A}$ the vector potential. Taking the curl of equations (17) yields

$$\mathrm{curl}\,\boldsymbol{p} - m_X\,\mathrm{curl}\,\boldsymbol{v} - 2\pi\hbar\; n_\phi \boldsymbol{\nu} - \frac{e_X}{c}\boldsymbol{B}, \tag{18}$$

where n_ϕ stands for the number density of vortices, $\boldsymbol{\nu} = (\mathrm{curl}\,\boldsymbol{v})\,/|\,\mathrm{curl}\,\boldsymbol{v}|$ and $\boldsymbol{B} = \mathrm{curl}\,\boldsymbol{A}$. The Maxwell equation for the magnetic field is

$$\mathrm{curl}\,\boldsymbol{B} = \frac{4\pi}{c} e n_c \boldsymbol{v}, \tag{19}$$

where n_c is number density of charge carriers. Taking the curl of Eq. (19), we find the analogue of the London equation

$$\mathrm{curl}\,\mathrm{curl}\,\boldsymbol{B} \;=\; \lambda^{-2}\left(\phi^* n_\phi \boldsymbol{\nu} - \boldsymbol{B}\right), \tag{20}$$

where

$$\lambda = \left[\frac{4\pi e e_X}{m_X c^2} n_c\right]^{1/2}, \quad \phi^* = \frac{2\pi\hbar c}{e_X}. \tag{21}$$

In the case of interest, we have $e_X = e$ and $m_X = 2m^*$, where m^* is a nucleon effective mass. Therefore,

$$\lambda = \left[\frac{\pi e^2}{m^* c^2}(1-\alpha)\right]^{1/2}, \quad \phi^* = \frac{2\pi\hbar c}{e} = 2\phi_0, \tag{22}$$

where $n_c \equiv \rho_p = (1-\alpha)\rho/2$ is the density of charge carriers. It seen from (22) that the quantum vortices present in the neutron-proton condensate carry twice the quantum of magnetic flux.

The values of λ and corresponding values of κ are given in Table 1 for representative points in the the coupling regimes WCR, ICR, and SCR. The gap value $\Delta = 1.27$ MeV found for the LOFF phase is much smaller than its corresponding value in the BCS state. Accordingly, to obtain a value for the coherence length in the LOFF phase, we have rescaled the $\xi_{\rm rms}$ value appearing in the table for the WCR case by the ratio of the pairing gap in the BCS state to that in the LOFF state, which is 9.39/1.27. This yields $\xi_{\rm rms}^{\rm LOFF} \simeq 23.4$ fm, which implies $\kappa_{\rm rms}^{\rm LOFF} = 0.8$. Since the GL parameter is very close to the critical value $1/\sqrt{2}$ this particular phase is close to the transition to a type-I superconducting state for values of the gap $\Delta \leq 1$ MeV.

4. Conclusions and perspectives

Low-density nuclear and neutron matter feature a rich phase diagrams in the presence of isospin asymmetry or spin polarization respectively. These phase diagrams may contain a number of phases: the translationally and rotationally symmetric, but spin/isospin polarized BCS phase, the BEC phase containing quasi-deuterons, the current-carrying LOFF phase, and the phase-separated phase.

In this contribution we have concentrated on the electrodynamics of these phases across the phase diagram derived in Refs. [12, 13]. We have evaluated the penetration depth of the magnetic field of nucleonic matter in various regions of the phase diagram and combined it with the microscopically computed coherence length to obtain the Ginzburg-Landau parameter. We find that in the entire region of the phase diagram considered here (with few exceptions, see below), the condensed phases are type-II superconductors which must form Abrikosov quantum vortices. However, because the neutron-proton Cooper pairs carry only a unit of charge, these vortices will carry twice the quantum of circulation ϕ_0. This is the main feature distinguishing the neutron-proton condensate from ordinary "condensed matter" superconductors and the proton superfluid in the core of a neutron star. In addition, we find that the LOFF phase is at the point of transition between the type-I and type-II superconducting states for gap values of the order of MeV. Because sufficiently large asymmetries will suppress the gap to values $\Delta \leq 1$ MeV, it is clear that this phase may eventually turn into a type-I superconductor. This argument applies to practically any phase with asymmetry, as $\xi \to \infty$ when $\Delta(\alpha) \to 0$ and therefore one finds $\kappa \to 0$. However, the relevant values of Δ might be extremely small for other phases.

In this contribution we have followed an approach based on phenomenological Ginzburg-Landau theory. It would be interesting and informative to extend our study of the electrodynamics of neutron-proton condensates to the microscopic level by evaluating the response functions of the condensates to electromagnetic fields (photon) in terms of the Green's functions introduced in Sec. 2.

Acknowledgments

We are grateful to Martin Stein and Xu-Guang Huang for collaboration on the topics covered in Sec. 2. AS is supported by the Deutsche Forschungsgemeinschaft (Grant No. SE 1836/4-1). JWC acknowledges research support from the McDonnell Center for the Space Sciences and expresses his thanks to Professor José Luís da Silva and his colleagues in the Centro de Investigação em Matemática e Aplicações during extended residence at the University of Madeira.

References

[1] Nozières P and Schmitt-Rink S 1985 *J. Low Temp. Phys.* **59** 195–211
[2] Alm T, Friman B L, Röpke G and Schulz H 1993 *Nuclear Phys. A* **551** 45–53
[3] Baldo M, Lombardo U and Schuck P 1995 *Phys. Rev. C* **52** 975–985
[4] Lombardo U, Nozières P, Schuck P, Schulze H J and Sedrakian A 2001 *Phys. Rev. C* **64** 064314
[5] Sedrakian A and Clark J W 2006 *Phys. Rev. C* **73** 035803
[6] Mao S, Huang X and Zhuang P 2009 *Phys. Rev. C* **79** 034304
[7] Huang X G 2010 *Phys. Rev. C* **81** 034007
[8] Isaule F, Arellano H F and Rios A 2016 *Phys. Rev. C* **94** 034004
[9] Rubtsova O A, Kukulin V I, Pomerantsev V N and Müther H 2017 *Phys. Rev. C* **96** 034327
[10] Fan H H, Krotscheck E and Clark J W 2017 *J. Low. Temp. Phys.* **189** 470–494
[11] Sedrakian A and Lombardo U 2000 *Phys. Rev. Lett.* **84** 602–605
[12] Stein M, Huang X G, Sedrakian A and Clark J W 2012 *Phys. Rev. C* **86** 062801
[13] Stein M, Sedrakian A, Huang X G and Clark J W 2014 *Phys. Rev. C* **90** 065804
[14] Stein M, Sedrakian A, Huang X G and Clark J W 2016 *Phys. Rev. C* **93** 015802
[15] Stein M, Sedrakian A, Huang X G, Clark J W and Röpke G 2014 *J. Phys. Conf. Series* **496** 012008

[16] Clark J W, Sedrakian A, Stein M, Huang X G, Khodel V A, Shaginyan V R and Zverev M V 2016 *J. Phys. Conf. Series* **702** 012012
[17] Buckley K B, Metlitski M A and Zhitnitsky A R 2004 *Phys. Rev. Lett.* **92** 151102
[18] Alford M, Good G and Reddy S 2005 *Phys. Rev. C* **72** 055801
[19] Sedrakian A 2005 *Phys. Rev. D* **71** 083003
[20] Alford M G and Good G 2008 *Phys. Rev. B* **78** 024510
[21] Sinha M and Sedrakian A 2015 *Phys. Rev. C* **91** 035805
[22] Haber A and Schmitt A 2017 *Phys. Rev. D* **95** 116016
[23] Haskell B and Sedrakian A 2017 *ArXiv e-prints* (*Preprint* 1709.10340)

IOP Conf. Series: Journal of Physics: Conf. Series **1041** (2018) 012009 doi:10.1088/1742-6596/1041/1/012009

Static density response of one-dimensional soft bosons across the clustering transition

M Teruzzi[1,2], D Pini[2], S Rossotti[2], D E Galli[2], G Bertaina[2]
[1] International School for Advanced Studies (SISSA), Via Bonomea 265, I-34136 Trieste, Italy
[2] Dipartimento di Fisica, Università degli Studi di Milano, via Celoria 16, I-20133 Milano, Italy

E-mail: gianluca.bertaina@unimi.it

Abstract. One-dimensional bosons interacting via a soft-shoulder potential are investigated at zero temperature. The flatness of the potential at short distances introduces a typical length, such that, at relatively high densities and sufficiently strong interactions, clusters are formed, even in the presence of a completely repulsive potential. We evaluate the static density response function of this system across the transition from the liquid to the cluster liquid phases. Such quantity reveals the density modulations induced by a weak periodic external potential, and is maximal at the clustering wavevector. It is known that this response function is proportional to the static structure factor in the classical regime at high temperature, while for this zero-temperature quantum systems, we extract it from the dynamical structure factor evaluated with quantum Monte Carlo methods.

1. Introduction

Ensembles of particles, interacting with potentials that are flat at short distances, have the possibility of clustering, at sufficiently high particle densities, even if the interaction is completely repulsive. This is a subject of intense study in classical soft-matter physics, where such kind of potentials may describe an effective interaction between polymers [1–4]. Recently, a great deal of interest has arisen in the quantum statistical mechanics of particles interacting with finite-range soft potentials, and the related quantum cluster phases. Supersolid behavior, characterized by the coexistence of crystal and superfluid order, has been investigated for bosons [5–9], and a breakdown of Fermi liquid theory has been predicted for fermions [10]. In one dimension (1D), cluster Luttinger liquids (CLL) have been proposed on a lattice [11, 12] and in the continuum [13]. The best candidates, for the experimental realization of such novel quantum phases, seem to be ultracold Rydberg gases [14], which are atoms in highly-excited electronic states, and especially ensembles of *dressed* Rydberg atoms, which are superpositions of the ground state and the above mentioned excited states, coupled via a Rabi process. Their effective interaction is a soft-shoulder potential, with a flat repulsive core of size R_c, related to the highly excited orbital, and a repulsive van-der-Waals tail [5, 6, 15–18]. Experiments are progressively increasing the coherence time of such systems [19–21].

In [13], we have characterized the $T = 0$ quantum phase transition [22] from a liquid to a cluster phase in a bosonic system, at a suitable density, by means of quantum Monte Carlo methods. In particular, we have evaluated the dynamical structure factor $S(q, \omega)$, which offers a picture of how the system responds to small dynamical external density perturbations, such as those coming from scattering experiments, providing the dispersion relation of collective modes.

IOP Conf. Series: Journal of Physics: Conf. Series **1041** (2018) 012009 doi:10.1088/1742-6596/1041/1/012009

In this article, we instead focus on the static response function $\chi(q)$, which provides information on how the system responds to a static periodic potential. This is relevant for trapped quantum gases, where external perturbations can be modulated quite easily. We also briefly describe some of the features of the liquid to cluster-liquid transition.

2. Methods

We study a system of N bosons in 1D at linear particle density n, governed by the following Hamiltonian in the continuum:

$$H = -\frac{\hbar^2}{2m}\sum_i^N \frac{\partial^2}{\partial x_i^2} + \sum_{i<j} \frac{V_0}{r_{ij}^6 + R_c^6}\,, \tag{1}$$

where x_i are the particle coordinates, $r_{ij} = |x_i - x_j|$ the distances, m is the mass and V_0 and R_c are the strength and the radius of the soft-shoulder potential $V(r)$. A possible realization of this potential in 1D is given by ultracold gases in optical elongated traps. The transverse degrees of freedom would be frozen in their ground state, due to low temperature and strong trapping. The physical hard-core radius c of the core electrons orbitals could be safely ignored, because of integration along the transverse direction, provided the three-dimensional density is small with respect to $1/c^3$. Anyway, the addition of a sufficiently small hard core $c \ll R_c$ to $V(r)$ should not qualitatively change the conclusions of this work, as discussed, in the higher dimensional case, in [23].

The low-energy physics of 1D quantum fluids is quite generally described by the Luttinger liquid (LL) paradigm [24], characterized by a phononic dispersion law $\varepsilon(q) = \hbar v q$. For Galilean invariant systems, sound velocity v is related to the Luttinger parameter K_L by $v = \hbar n\pi/(mK_L)$ [25]. In order to study the excitations at higher energies and momenta, typically numerical methods are required, and we analyze the system using the path integral ground state (PIGS) quantum Monte Carlo method [26, 27], which projects a trial wavefunction in imaginary time to the ground state. We simulate up to $N = 200$ particles in a segment of length $L = N/n$, using periodic boundary conditions (PBC) (see [13]). We have thus access to the exact static structure factor $S(q)$ of the system. Moreover, evaluation of the dynamical structure factor $S(q,\omega) = \int dt \frac{e^{i\omega t}}{2\pi N}\langle e^{itH/\hbar}\rho_q e^{-itH/\hbar}\rho_{-q}\rangle$, where ρ_q is the density operator in momentum space, is provided via analytic continuation, through the genetic inversion via falsification of theories algorithm [28–31].

3. Clustering and phase diagram

It is convenient to introduce the dimensionless density $\rho = nR_c$ and strength $U = V_0/(E_cR_c^6)$, where $E_c = \hbar^2/mR_c^2$. The Fourier transform $\tilde{V}(q)$ of the potential has the following form:

$$\tilde{V}(q) = E_c\frac{U}{3}\sqrt{\frac{\pi}{2}}e^{-|q|R_c/2}\left[\cos\left(\frac{\sqrt{3}}{2}|q|R_c\right) + e^{-|q|R_c/2} + \sqrt{3}\sin\left(\frac{\sqrt{3}}{2}|q|R_c\right)\right]. \tag{2}$$

It features a global minimum, found numerically at $q_c \simeq 4.3$, for which $\tilde{V}(q_c) < 0$, corresponding to a typical length $b_c = 2\pi/q_c \simeq 1.46$ (here and in the following, when not otherwise specified, lengths and wavevectors are in units of R_c and $1/R_c$, respectively).

In the classical case it has been shown that, even for a completely repulsive potential, such a feature favors the formation of clusters at a mutual distance $\sim b_c$ which, at suitably low temperature T or high density ρ, arrange into an ordered configuration, thereby forming a cluster crystal with periodicity determined by b_c itself [1, 2]. To be more specific, in order to have stable clusters, a positive clustering harmonic constant (see [13, 32]) is needed. For such

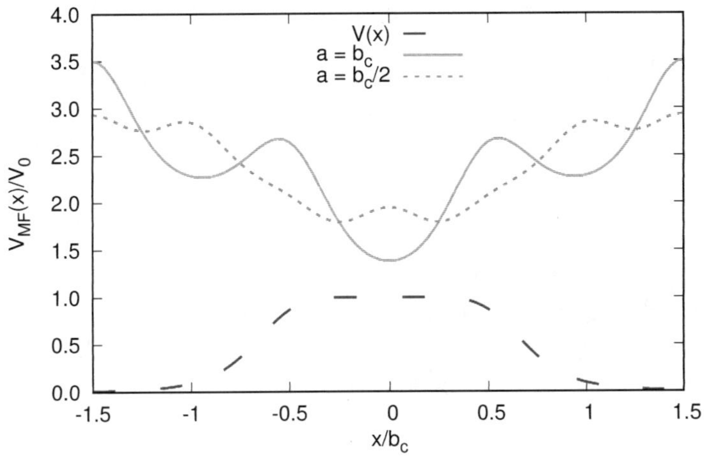

Figure 1. Mean-field potentials experienced by a reference particle, at $\rho \simeq 1.37$, if all the others are placed in a lattice of spacing $a = b_c$ (two particles per site) or $a = b_c/2$ (one particle per site). The original soft-shoulder potential is also shown as a reference.

condition to be satisfied it is necessary (but not sufficient) for $\tilde{V}(q)$ to take negative values for some q. For the one-dimensional system considered here, this ordered phase is most likely prevented by thermal fluctuations at all temperatures $T > 0$ [33, 34], but, classically, it still occurs at $T = 0$ for any positive value of U. Indeed, it has been argued that, on increasing ρ, one will witness a virtually infinite sequence of first-order transitions connecting phases with different numbers of particles per cluster [32, 34].

Let us focus on the density $\rho_2 = 2/b_c \simeq 1.37$. Classically, at $T = 0$, such a density accommodates a crystal of two-particles clusters. This can be easily understood by considering the mean-field potential experienced by a particle, when all the others are placed in couples on a perfect lattice of spacing $a = b_c$, except for a companion particle, which is placed at $x = 0$. The resulting potential (see solid line in Fig.1) manifests a clear harmonic minimum at $x = 0$. Notice also that secondary minima are present at multiples of b_c, meaning that, upon increasing kinetic energy, hopping to nearby clusters is initially more preferable than melting of the clusters. Conversely, if one forces the other particles to sit on lattice sites with spacing $a = b_c/2$, the mean-field potential displays a double-well structure, close to $x = 0$ (dotted line), indicating that such a perfect lattice, with one particle per site, would not be an equilibrium configuration.

In the quantum case, the scenario is further enriched, since zero-point kinetic energy is present even at $T = 0$, introducing two main consequences: the cluster phase is not a crystal anymore, but a liquid of clusters, and clusters can also coherently delocalize towards a standard Luttinger liquid. As a consequence, a quantum phase transition occurs between the CLL and a LL without cluster order, at a (density-dependent) value of U (Fig. 2). This transition has been studied in a related 1D lattice model [11, 12], while we have demonstrated that, at the density $\rho = \rho_2$ in the continuum, it occurs at $U = U_c \simeq 18$ and is in the 2D Ising universality class [13].

In the dilute limit, the only relevant parameter is the scattering length, which we have calculated for the soft-shoulder potential in 1D [35]. At $\rho = \rho_2$, on the contrary, the details of $V(r)$ become relevant. In the LL regime, a Bogoliubov picture turns out to be valid [13], providing the following dispersion relation for the main excited mode

$$\varepsilon_B(q) = \sqrt{\varepsilon_0(q) \left[\varepsilon_0(q) + 2\rho\tilde{V}(q)\right]}\,, \tag{3}$$

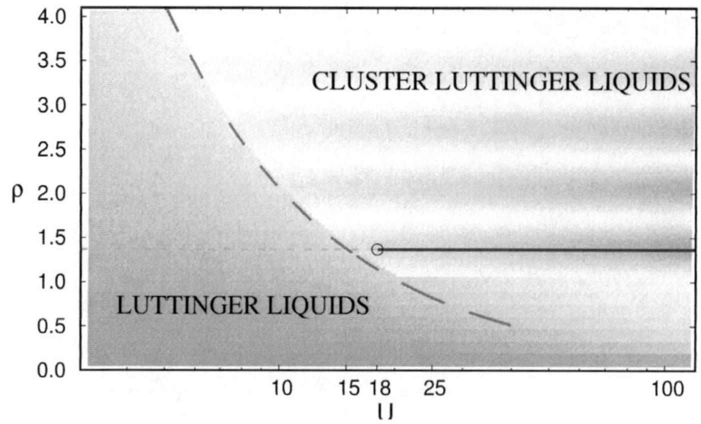

Figure 2. Phase diagram (linear-log scale). A symbol indicates the critical point between the LL (dashed line) and CLL (solid line) phases for density $\rho = 1.37$ [13], commensurate to two-particle clusters. The long-dashed line corresponds to the softening of the Bogoliubov roton.

where $\varepsilon_0(q) = \hbar^2 q^2/2m$ is the free particle dispersion. Interestingly, the functional form of the Bogoliubov excitation allows for a minimum of $\varepsilon_B(q)$ at $q \simeq q_c$, provided the combination $\alpha = \rho U$ is sufficiently large. The dynamical structure factors, evaluated in [13], indeed show very clearly that a roton minimum is prominent in the LL regime. When $\alpha = 20.65$, the Bogoliubov roton softens, and this mean-field picture breaks down. This can be used as an approximate boundary between the LL and CLL phases (long-dashed line in Fig. 2), although microscopic calculations have to be performed to determine the precise position (see, for example, the circle in Fig. 2, at $\rho = \rho_2$). It is still an open issue the precise determination of the transition between different quantum cluster phases upon increasing ρ.

4. Static density response function

The static density response function is introduced, in linear response theory, as the coefficient of proportionality between a weak static periodic perturbation $\tilde{v}(q)$ and the produced density fluctuation $\delta\rho(q) \simeq \chi(q)\tilde{v}(q)$, with respect to the equilibrium homogeneous system [36, 37]. At zero temperature, it can be computed by carefully evaluating energy differences between perturbed and unperturbed systems [36] or from the first negative moment of $S(q,\omega)$ [38]:

$$\chi(q) = -\frac{2\rho}{\hbar}\int_0^\infty d\omega\,\frac{S(q,\omega)}{\omega}\,. \tag{4}$$

The previous equation is the $T \to 0$ limit of the more general fluctuation-dissipation relation

$$\chi(q) = -\frac{2\rho}{\hbar}\int_0^\infty d\omega\,\frac{(1-e^{-\beta\hbar\omega})S(q,\omega)}{\omega}\,, \tag{5}$$

with $\beta = 1/k_{\rm B}T$, $k_{\rm B}$ being the Boltzmann constant. It is straightforward for us to evaluate Eq. (4), by integrating $S(q,\omega)$ as obtained in [13]. Our results are shown in Fig. 3. We compare $\chi(q)$ from Eq. (4) to the Feynman approximation of the same quantity:

$$\chi_{FA}(q) = -2\rho\frac{S(q)^2}{\varepsilon_0(q)}\,. \tag{6}$$

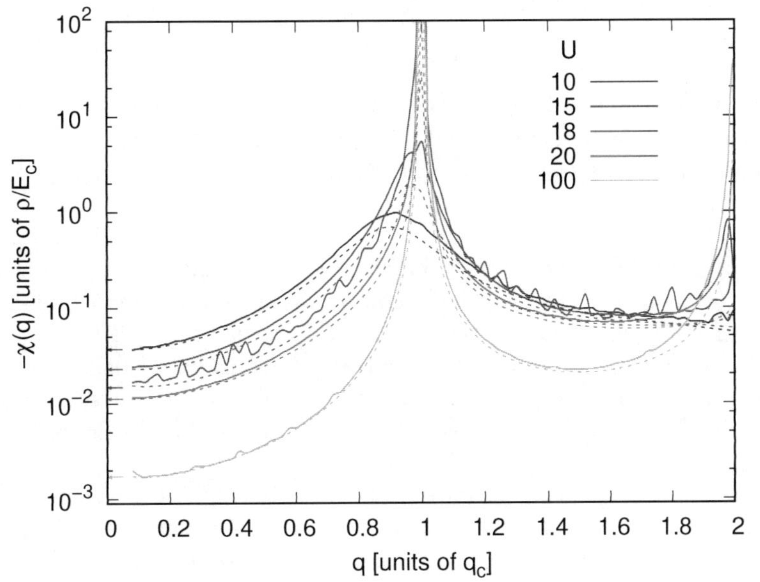

Figure 3. Static density response function $\chi(q)$ (with a negative sign), evaluated as (solid lines) the first negative moment of the dynamical structure factor, compared to Feynman single-mode approximation $\chi_{FA}(q)$ (dashed lines), for various U across the transition at $\rho \simeq 1.37$. The arrows indicate the compressibility as extracted from the slope of the static structure factor at at $q \to 0$. Errorbars (not shown for clarity) are of the order of the fluctuations of the plotted quantities, and are large only for $U = 18$, which approximately corresponds to the transition point and thus suffers more from the finite length of the simulations.

The latter assumes that the density fluctuations spectrum is exhausted by a single mode of dispersion $\varepsilon_{FA}(q)$, so that the zeroth

$$S(q) = \int_0^\infty d\omega S(q,\omega) \tag{7}$$

and first

$$\varepsilon_0(q)/\hbar = \int_0^\infty d\omega S(q,\omega)\omega \tag{8}$$

sum rules imply $\varepsilon_{FA}(q) = \varepsilon_0(q)/S(q)$ and $S_{FA}(q,\omega) = S(q)\delta(\omega - \varepsilon_{FA}(q)/\hbar)$, allowing for an estimation of various quantities provided the knowledge of only the static structure factor [39]. We observe that, qualitatively, $\chi_{FA}(q)$ grasps the main features of $\chi(q)$.

It is worth recalling that, in the classical case, $\chi(q)$ is simply proportional to $S(q)$ via the relation $\chi(q) = -\beta\rho S(q)$, which is recovered from the finite-temperature quantum expression of $\chi(q)$, Eq. (5), in the classical limit $\beta\hbar\omega \to 0$. For the classical fluid, the connection between cluster formation and the occurrence of the global minimum of $\tilde{V}(q)$ at $q_c \neq 0$ is then brought forward within a simple mean-field approximation for $S(q)$, according to which $\chi(q) = -\beta\rho/[1+\beta\rho\tilde{V}(q)]$. The minimum of $\tilde{V}(q)$ then causes a peak at $q = q_c$ in both $S(q)$ and $\chi(q)$, signaling the attitude of the system to form density modulations over lengthscales $\sim b_c$. The above mean-field expression refers to the classical fluid phase and is not valid in the limit $T \to 0$. Moreover, in this limit the classical and quantum expressions of $\chi(q)$ differ markedly and, as shown by Eq. (4), in the latter case the connection between $\chi(q)$ and $S(q)$ gets lost.

In particular, the quantum $S(q)$ becomes linear in q at small q, whereas $\chi(q)$ tends to a non-vanishing value which, as in the classical case, is proportional to the isothermal compressibility of the system $\chi(q)\to -\rho/mv^2$.
Notwithstanding these differences, Fig. 3 clearly shows that, even in the quantum case, $\chi(q)$ still displays a peak at $q \sim q_c$ which becomes more and more pronounced on increasing U, similarly to what predicted by the above mean-field approximation for the classical $\chi(q)$ on increasing β.

A comment on finite-size effects is in order. Strictly speaking, $\chi(q)$ (and analogously $S(q)$) may diverge, at a specific wavevector, only in the $N \to \infty$ limit. The number $N = 200$ allows us to see a clearly different behavior between $U < U_c$ and $U \geq U_c$: namely, before the transition, the maximum of $\chi(q)$ saturates at a finite value (liquid behavior), while, for stronger interaction, this maximum diverges. It can be shown that such a divergence is sub-linear with the number of particles (quasi-solid behavior, see [13] for a detailed analysis of finite-size effects of $S(q)$). Moreover, the number of particles, with PBC, directly sets the momentum resolution $\delta q = 2\pi\rho/N$. Finally, since we want to reduce the creation of defects in the two-particles cluster phase, we only simulate an even number of bosons.

The striking feature of this peak of $\chi(q)$ (and $S(q)$) is that its position is determined by the shape of the interaction potential, and not by the mean interparticle distance $1/\rho$, which, for $\rho = \rho_2$, would correspond to $q = 2q_c$. The static response function we have evaluated thus indicates a very strong sensitivity of the system to external potentials with modulation of wavevector $\sim q_c$, as expected, both in the liquid regime and especially in the cluster regime. By explicitly breaking translational invariance, such perturbations would induce the formation of a true crystal, analog to the superfluid to Mott transition of one-dimensional dilute Bose gases [40, 41].

We finally discuss the physical understanding of the crucial role of the shape of the potential in favoring clustering. Clearly, the property of the potential of having a negative minimum in its Fourier transform depends on the overall shape of $V(r)$, rather than on its strength V_0. A change in V_0 would merely amount to a rescaling of the Fourier transform of the interaction, which would affect the depth of its minimum, but would leave its position q_c unchanged.

In principle, the occurrence of a negative minimum in $\tilde{V}(q)$ may be caused by a number of different features of the interaction. For the potential considered here, it stems from $V(r)$ being rather flat at short distances, while decaying rapidly at longer distances. How such a property may favor cluster formation can be explained heuristically by mean-field arguments similar to that put forward in the discussion of Fig. 1, such as those presented for a classical system in [42]. In particular, let us consider a one-dimensional regular array of particles with nearest-neighbor distance d, interacting with each other via a repulsive, bounded potential $V(r)$, and let us determine the overall potential $W(r)$ felt by a further test particle inserted into the system for small enough d. If $V(r)$ is, say, a Gaussian (whose Fourier transform clearly does not have a negative minimum), then it is found that $W(r)$ has minima in between the particles of the array, suggesting that the preferred position of the additional particle would destroy the periodicity. However, if $V(r)$ is made flatter at small r and steeper at larger r, then $W(r)$ has minima on the top of the particles of the array: this would drive the test particle on top of one of these particles, thereby preserving the periodicity by forming a cluster. In such a process, the optimal value of d depends on the detailed shape of $V(r)$, whereas the strength of $V(r)$ does not play any role. An extreme (admittedly artificial) example of the latter instance is that of a square-shoulder $V(r)$: when the density of the array reaches the close-packing value such that d is equal to the shoulder width, placing an additional particle in between those of the array will create two overlaps, whereas putting it on top of one of them will create just one.

We may then say that for this class of potentials, the interaction energy price one has to pay for clustering is more than compensated by the ensuing reduction of the repulsion between particles on neighboring sites. The role of the strength V_0 is to tune the relative dominance of potential energy with respect to kinetic energy, and is thus particularly relevant in the 1D quantum case, where it triggers the transition to the liquid phase [13].

Acknowledgments
We acknowledge useful discussions with A. Parola. We acknowledge the CINECA awards IscraC-SOFTDYN-2015 and IscraC-CLUDYN-2017 for the availability of high performance computing resources and support. M. T. also acknowledges funding from the European Research Council (ERC) under the European Union's Horizon 2020 research and innovation programme (grant agreement No. 320796-MODPHYSFRICT).

References

[1] Likos C N, Lang A, Watzlawek M and Löwen H 2001 *Phys. Rev.* E **63** 031206
[2] Mladek B M, Gottwald D, Kahl G, Neumann M and Likos C N 2006 *Phys. Rev. Lett.* **96** 045701
[3] Mladek B M, Kahl G and Likos C N 2008 *Phys. Rev. Lett.* **100** 028301
[4] Lenz D A, Blaak R, Likos C N and Mladek B M 2012 *Phys. Rev. Lett.* **109** 228301
[5] Henkel N, Nath R and Pohl T 2010 *Phys. Rev. Lett.* **104** 195302
[6] Cinti F, Macrì T, Lechner W, Pupillo G and Pohl T 2014 *Nat. Commun.* **5** 3235
[7] Saccani S, Moroni S and Boninsegni M 2012 *Phys. Rev. Lett.* **108** 175301
[8] Ancilotto F, Rossi M and Toigo F 2013 *Phys. Rev.* A **88** 033618
[9] Angelone A, Mezzacapo F and Pupillo G 2016 *Phys. Rev. Lett.* **116** 135303
[10] Li W H, Hsieh T C, Mou C Y and Wang D W 2016 *Phys. Rev. Lett.* **117** 035301
[11] Mattioli M, Dalmonte M, Lechner W and Pupillo G 2013 *Phys. Rev. Lett.* **111** 165302
[12] Dalmonte M, Lechner W, Cai Z, Mattioli M, Läuchli A M and Pupillo G 2015 *Phys. Rev.* B **92** 045106
[13] Rossotti S, Teruzzi M, Pini D, Galli D E and Bertaina G 2017 *Phys. Rev. Lett.* **119** 215301
[14] Löw R, Weimer H, Nipper J, Balewski J B, Butscher B, Büchler H P and Pfau T 2012 *J. Phys.* B **45** 113001
[15] Pupillo G, Micheli A, Boninsegni M, Lesanovsky I and Zoller P 2010 *Phys. Rev. Lett.* **104** 223002
[16] Balewski J B, Krupp A T, Gaj A, Hofferberth S, Löw R and Pfau T 2014 *New J. Phys.* **16** 063012
[17] Macrì T, Saccani S and Cinti F 2014 *J. Low Temp. Phys.* **177** 59–71
[18] Płodzień M, Lochead G, de Hond J, van Druten N J and Kokkelmans S 2017 *Phys. Rev.* A **95** 043606
[19] Jau Y Y, Hankin A M, Keating T, Deutsch I H and Biedermann G W 2016 *Nat. Phys.* **12** 71–74
[20] Zeiher J, van Bijnen R, Schauß P, Hild S, Choi J Y, Pohl T, Bloch I and Gross C 2016 *Nat. Phys.* **12** 1095–1099
[21] Zeiher J, Choi J Y, Rubio-Abadal A, Pohl T, van Bijnen R, Bloch I and Gross C 2017 *arXiv:1705.08372*
[22] Sachdev S 2000 *Quantum Phase Transitions* (Cambridge: Cambridge University Press)
[23] Boninsegni M 2016 *J. Low Temp. Phys.* **184** 1071–1079
[24] Giamarchi T 2003 *Quantum Physics in One Dimension* (Oxford University Press)
[25] Haldane F D M 1981 *Phys. Rev. Lett.* **47** 1840–1843
[26] Sarsa A, Schmidt K E and Magro W R 2000 *J. Chem. Phys.* **113** 1366–1366
[27] Rossi M, Nava M, Reatto L and Galli D E 2009 *J. Chem. Phys.* **131** 154108
[28] Vitali E, Rossi M, Reatto L and Galli D E 2010 *Phys. Rev.* B **82** 174510
[29] Bertaina G, Motta M, Rossi M, Vitali E and Galli D E 2016 *Phys. Rev. Lett.* **116** 135302
[30] Motta M, Vitali E, Rossi M, Galli D E and Bertaina G 2016 *Phys. Rev.* A **94** 043627
[31] Bertaina G, Galli D E and Vitali E 2017 *Adv. Phys. X* **2** 302–323
[32] Neuhaus T and Likos C N 2011 *J. Phys.: Condens. Matter* **23** 234112
[33] Prestipino S 2014 *Phys. Rev.* E **90** 042306
[34] Prestipino S, Gazzillo D and Tasinato N 2015 *Phys. Rev.* E **92** 022138
[35] Teruzzi M, Galli D E and Bertaina G 2017 *J. Low Temp. Phys.* **187** 719–726
[36] Moroni S, Ceperley D M and Senatore G 1995 *Phys. Rev. Lett.* **75** 689–692
[37] Senatore G, Moroni S and Ceperley D M 1998 *Quantum Monte Carlo Methods in Physics and Chemistry* Nato Science Series C: ed Nightingale M and Umrigar C (Springer Netherlands)
[38] Motta M, Bertaina G, Vitali E, Galli D E and Rossi M 2017 *J. Low Temp. Phys.* **187** 419–426
[39] Hall D and Feenberg E 1971 *Ann. Phys.* **63** 335–360
[40] Stöferle T, Moritz H, Schori C, Köhl M and Esslinger T 2004 *Phys. Rev. Lett.* **92** 130403
[41] Fertig C D, O'Hara K M, Huckans J H, Rolston S L, Phillips W D and Porto J V 2005 *Phys. Rev. Lett.* **94** 120403
[42] Mladek B M, Charbonneau P, Likos C N, Frenkel D and Kahl G 2008 *J. Phys.: Condens. Matter* **20** 494245

An analysis of variational wave function for the pairing problem in strongly correlated system

H.-H. Fan†‡ **and E. Krotscheck**†‡

†Department of Physics, University at Buffalo, SUNY Buffalo NY 14260
‡Institut für Theoretische Physik, Johannes Kepler Universität, A 4040 Linz, Austria

E-mail: `eckhardk@buffalo.edu`

Abstract.
We report a theoretical analysis of variational wave functions for the BCS pairing problem. Starting with a Jastrow–Feenberg (or, in a more recent language "fixed–node") wave function for the superfluid state, we develop the full optimized Fermi-Hypernetted Chain (FHNC-EL) equations which sum a local approximation of the parquet–diagrams. Close examination of the procedure reveals that it is essential to go *beyond* the usual Jastrow–Feenberg approximation to guarantee the correct stability range.

1. Motivation

Jastrow–Feenberg theory has been established as a very effective method for calculating ground state properties of strongly interacting systems. We assume a non-relativistic many-body Hamiltonian

$$H = -\sum_i \frac{\hbar^2}{2m}\nabla_i^2 + \sum_{i<j} V(i,j)\,, \tag{1}$$

or, in second quantized form,

$$\hat{H} = \sum_{\alpha,\beta}\langle\alpha|T|\beta\rangle a_\alpha^\dagger a_\beta + \frac{1}{2}\sum_{\alpha,\beta,\gamma,\delta}\langle\alpha\beta|V|\gamma\delta\rangle a_\alpha^\dagger a_\beta^\dagger a_\delta a_\gamma\,. \tag{2}$$

The method starts with an *ansatz* for the wave function, [1]

$$\Psi_0(\mathbf{r}_1,\ldots,\mathbf{r}_N) = \frac{1}{\sqrt{I_{\mathbf{o},\mathbf{o}}}}F_N(\mathbf{r}_1,\ldots,\mathbf{r}_N)\Phi_0(\mathbf{r}_1,\ldots,\mathbf{r}_N), \tag{3}$$

$$F_N(\mathbf{r}_1,\ldots,\mathbf{r}_N) = \exp\frac{1}{2}\left[\sum_{i<j}u_2(\mathbf{r}_i,\mathbf{r}_j)+\cdots+\sum_{i_1<\ldots<i_n}u_n(\mathbf{r}_{i_1},..,\mathbf{r}_{i_n})+\cdot\cdot\right], \tag{4}$$

where $I_{\mathbf{o},\mathbf{o}} = \left\langle\Phi_0|F_N^\dagger F_N|\Phi_0\right\rangle$ is the normalization constant. Here $\Phi_0(\mathbf{r}_1,\ldots,\mathbf{r}_N)$ denotes a model state, which for normal Fermi systems is a Slater-determinant, and F_N is an N-body correlation operator written in general form, but to be truncated at the two-body u_2 term in a standard Jastrow calculation. The computationally most effective way to deal with expectation

IOP Conf. Series: Journal of Physics: Conf. Series **1041** (2018) 012010 doi:10.1088/1742-6596/1041/1/012010

values of correlated wave functions of the type (4) are diagrammatic methods, specifically the optimized Euler-Lagrange Fermi-hypernetted chain (FHNC-EL) method, which is well suited for calculation of physically interesting quantities. These diagrammatic methods have been successfully applied to such highly correlated Fermi systems as ^{3}He at $T = 0$ [2]. We have shown in recent work [3] that even the simplest version of the FHNC-EL theory is accurate within better than one percent at densities less than 25% of the saturation density of liquid ^{3}He and similarly for nuclear systems [4]. It is one of the most attractive features of the FHNC-EL method that the Euler equation (See Eq. (9) below) has no solution if the assumed ground state is incorrect.

Following up on previous work [5, 6] and partly motivated by interest in the BCS-BEC crossover in cold gases (see Refs. 7 and 8 for review articles), we have recently examined pairing phenomena in both model Fermi systems [9] and neutron matter interacting via the $^1S_0$ components of the Reid soft-core V_6 [10] and the Argonne V_4' [11] two-nucleon interactions. These calculations were based on a generalization of the BCS wave function

$$\left|\mathrm{BCS}\right\rangle = \prod_{\mathbf{k}} \left[u_{\mathbf{k}} + v_{\mathbf{k}} a^\dagger_{\mathbf{k}\uparrow} a^\dagger_{-\mathbf{k}\downarrow}\right] \left|0\right\rangle \tag{5}$$

of the form

$$\left|\mathrm{CBCS}\right\rangle = \sum_{\mathbf{m},N} \left|\Psi^{(N)}_{\mathbf{m}}\right\rangle\left\langle \mathbf{m}^{(N)}\middle|\mathrm{BCS}\right\rangle \tag{6}$$

where the correlated states N-body $\left|\Psi^{(N)}_{\mathbf{m}}\right\rangle$ are

$$\left|\Psi^{(N)}_{\mathbf{m}}\right\rangle = \left[I^{(N)}_{\mathbf{m},\mathbf{m}}\right]^{-1/2} F_N \left|\mathbf{m}^{(N)}\right\rangle , \qquad I^{(N)}_{\mathbf{m},\mathbf{m}} \equiv \left\langle \mathbf{m}^{(N)}\middle|F_N^\dagger F_N\middle|\mathbf{m}^{(N)}\right\rangle , \tag{7}$$

Here the $\{\left|\mathbf{m}^{(N)}\right\rangle\}$ form complete sets of N-particle Slater determinants of single-particle orbitals. We have commented on alternative choices of the correlated BCS wave function in Ref. 9, the arguments do not need to be repeated here.

Physically interesting quantities like the free energy of the superfluid system

$$\left\langle H'\right\rangle_s = \frac{\left\langle \mathrm{CBCS}\middle|\hat{H}'\middle|\mathrm{CBCS}\right\rangle}{\left\langle \mathrm{CBCS}\middle|\mathrm{CBCS}\right\rangle} , \qquad \hat{H}' \equiv \hat{H} - \mu\hat{N} . \tag{8}$$

are then calculated by cluster expansion and resummation techniques; the correlation functions are determined by the variational principle

$$\frac{\delta\left\langle H'\right\rangle_s}{\delta u_n}(\mathbf{r}_1,\ldots,\mathbf{r}_n) = 0 . \tag{9}$$

The combination of the (Fermi-)hypernetted-chain summation technique for the energy with the optimization of the ground state correlations is equivalent to a local approximation of the so–called "parquet-diagrams" [12, 13, 14].

In Refs. 9 and 4 we have simplified the problem by expanding the free energy (8) in the *deviation* of the Bogoliubov amplitudes $u_{\mathbf{k}}$, $v_{\mathbf{k}}$ from their normal state values $u^{(0)}_{\mathbf{k}} = \theta(k - k_{\mathrm{F}})$, $v^{(0)}_{\mathbf{k}} = \theta(k_{\mathrm{F}} - k)$. In that case, the pair correlation functions $u_2(r)$ can be optimized for the normal system. With that, we have arrived at the energy expression of the superfluid state

$$\begin{aligned}\langle \hat{H}'\rangle_s &= H^{(N)}_{00} - \mu N + 2\sum_{\mathbf{k},|\mathbf{k}|>k_{\mathrm{F}}} v^2_{\mathbf{k}}(e_{\mathbf{k}} - \mu) - 2\sum_{\mathbf{k},|\mathbf{k}|<k_{\mathrm{F}}} u^2_{\mathbf{k}}(e_{\mathbf{k}} - \mu) \\ &\quad + \sum_{\mathbf{k},\mathbf{k}'} u_{\mathbf{k}} v_{\mathbf{k}} u_{\mathbf{k}'} v_{\mathbf{k}'} \mathcal{P}_{\mathbf{k}\mathbf{k}'} .\end{aligned} \tag{10}$$

IOP Conf. Series: Journal of Physics: Conf. Series **1041** (2018) 012010 doi:10.1088/1742-6596/1041/1/012010

Above, $H_{00}^{(N)}$ is the energy of the normal N-particle system, μ is the chemical potential. The $e_{\mathbf{k}}$ are the single particle energies derived in correlated basis function (CBF) theory [15], and the paring interaction has the form

$$\begin{aligned}
\mathcal{P}_{\mathbf{k}\mathbf{k}'} &= \mathcal{W}_{\mathbf{k}\mathbf{k}'} + (|e_{\mathbf{k}} - \mu| + |e_{\mathbf{k}'} - \mu|)\mathcal{N}_{\mathbf{k}\mathbf{k}'}\,, && (11)\\
\mathcal{W}_{\mathbf{k}\mathbf{k}'} &= \big\langle \mathbf{k}\uparrow, -\mathbf{k}\downarrow \big| \mathcal{W}(1,2) \big| \mathbf{k}'\uparrow, -\mathbf{k}'\downarrow \big\rangle_a\,, && (12)\\
\mathcal{N}_{\mathbf{k}\mathbf{k}'} &= \big\langle \mathbf{k}\uparrow, -\mathbf{k}\downarrow \big| \mathcal{N}(1,2) \big| \mathbf{k}'\uparrow, -\mathbf{k}'\downarrow \big\rangle_a\,. && (13)
\end{aligned}$$

where the effective interaction $\mathcal{W}(1,2)$ and the correlation corrections $\mathcal{N}(1,2)$ are then given by the compound-diagrammatic ingredients of the FHNC-EL method for off-diagonal quantities in CBF theory [15].

The Bogoliubov amplitudes $u_{\mathbf{k}}$, $v_{\mathbf{k}}$ are obtained in the standard way by variation of the energy expectation (10). This leads to the familiar gap equation

$$\Delta_{\mathbf{k}} = -\frac{1}{2}\sum_{\mathbf{k}'} \mathcal{P}_{\mathbf{k}\mathbf{k}'} \frac{\Delta_{\mathbf{k}'}}{\sqrt{(e_{\mathbf{k}'} - \mu)^2 + \Delta_{\mathbf{k}'}^2}}\,. \qquad (14)$$

In what follows, we shall denote the value of the gap function $\Delta_{\mathbf{k}}$ at the Fermi surface with $\Delta \equiv \Delta_{|\mathbf{k}|=k_F}$.

The conventional (*i.e.* "uncorrelated" or "mean-field") BCS gap equation [16] is retrieved by replacing the effective interaction $\mathcal{P}_{\mathbf{k}\mathbf{k}'}$ by the pairing matrix of the bare interaction.

In the calculations of Refs. 4 and 9 we have, however, encountered two problems indicating that the "weak coupling approximation" can be problematic:

- We have examined in Ref. 9 a large array of model systems described by a Lennard-Jones potential

  $$V_{\rm LJ} = 4\epsilon\left[\left(\frac{\sigma}{r}\right)^{12} - \left(\frac{\sigma}{r}\right)^{6}\right] \qquad (15)$$

  and an attractive square well (SW) potential

  $$V_{\rm SW}(r) = \begin{cases} -\epsilon & \text{if} \quad r < \sigma\,, \\ 0 & \text{if} \quad r > \sigma\,. \end{cases} \qquad (16)$$

  Both potentials are parametrized by a characteristic length σ and the depth ϵ of the attractive well. For a large array of model systems with net attractive interactions (measured by a negative vacuum scattering length a_0) we found that the *in–medium scattering length* a diverges, as a function of interaction strength, well before the vacuum scattering length a_0 diverges, as shown in Fig. 1.
  The divergence of the in-medium scattering length appears to be an indication for the formation of bound dimers (in other words the BCS-BEC "crossover") which is enhanced by phonon exchange. Such a dimerized state is not described by the Jastrow–Feenberg wave function of a normal system. On the other hand, the simple BCS wave function (5) can describe the crossover [17, 18] in situations where many–body correlations are negligible. It is therefore expected that the correlated BCS state (6) can describe the system through the BCS-BEC crossover, taking many–body correlations like phonon–exchange properly into account.
- For systems like neutron matter that have no many-body bound state, many–body correlations are less important. Nevertheless we found in Ref. 4 – in agreement with very many previous calculations and also more recent fixed–node Monte-Carlo calculations

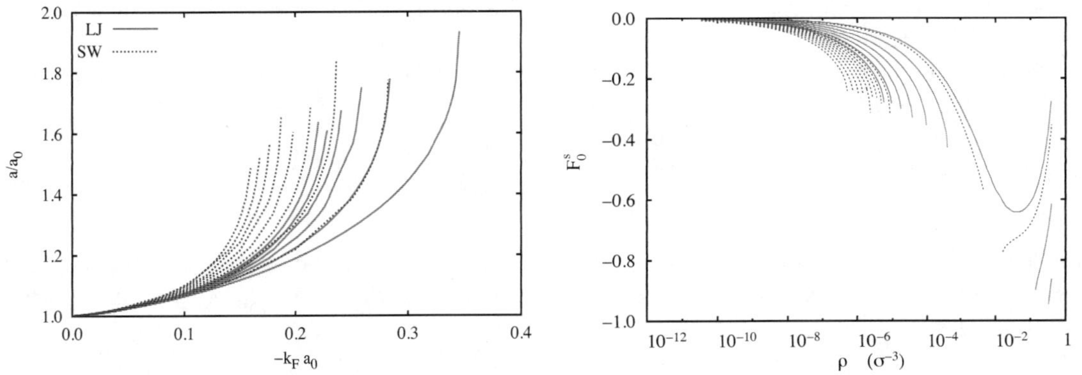

Figure 1. The left plot shows the ratio between the in-medium scattering length a and the vacuum scattering length a_0 as function of $-k_{\rm F}a_0$, for the Lennard-Jones (red lines) and the square-well potential (blue lines). The curves correspond to values of $a_0/\sigma = -0.5, -1.0, \ldots, -4.0$ (SW) and $a_0/\sigma = -1.5, -2.0, \ldots, -4.0$ (LJ), with the higher curves corresponding to larger $|a_0|$. The right plot shows the dependence of the Landau parameter F_0^s for the Lennard-Jones model of the interaction. The blue (dashed) lines correspond to coupling strengths $\epsilon = 7.0, 9.0, 9.2, 9.3, \ldots 9.9$. The curves for the interaction strengths $\epsilon = 7.0$ $(a_0/\sigma = -1.12)$ (blue, dashed), $\epsilon = 7.51$ $(a_0/\sigma = -1.5)$ and $\epsilon = 8.01$ $(a_0/\sigma = -2)$ (both red, solid) are discontinuous at high density. The curve that ends at the lowest density corresponds to the strongest interaction. From Ref. 9.

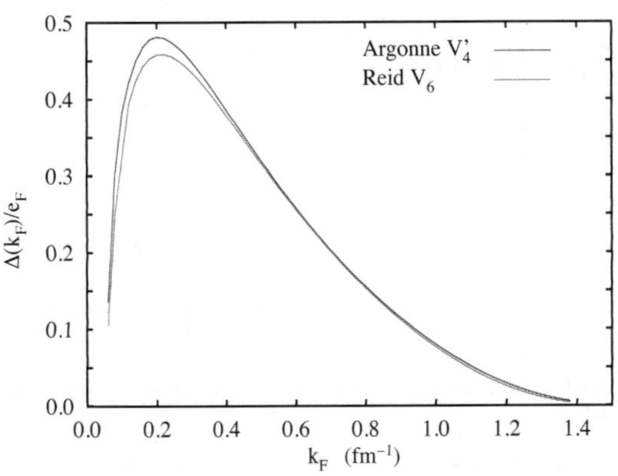

Figure 2. The figure shows the ratio $\Delta/e_{\rm F}$ in neutron matter for the Reid V_6 and the Argonne V_4' potentials. From Ref. 4.

[19, 20] at low densities, a pairing gap of the order of half of the Fermi energy $e_{\rm F}$, see Fig. 2. The derivations of the correlated BCS theory [9, 5, 6, 21] assume, on the other hand, that the paring gap is much smaller than the Fermi energy; the validity of the approximations leading to the formulation (10-13) must therefore be questioned.

2. BCS theory with local correlations

The divergence of the Euler equation observed in Ref. 9 – recall that we have assumed there a normal ground state and treated BCS correlations perturbatively – is therefore expected to be a signature of a crossover to a strongly dimerized state: The disappearance of solutions of the Euler equation for the *normal* system tells us that our assumption that the system is normal is incorrect, and the pairing is so strong that a perturbative expansion in terms of the deviation from the normal state as used previously is not legitimate.

IOP Conf. Series: Journal of Physics: Conf. Series **1041** (2018) 012010 doi:10.1088/1742-6596/1041/1/012010

To clarify the situation, we have developed the cluster expansion and resummation procedures for the fully correlated state (6).

The central quantity in the development of the method is the free energy (8) of the superfluid system which we can write as

$$\left\langle H'\right\rangle_s = \frac{\sum_{N,\mathbf{m},\mathbf{n}}\left\langle \mathrm{BCS}\middle|\mathbf{m}^{(N)}\right\rangle\left\langle \Psi^{(N)}_{\mathbf{m}}\middle|\hat{H}'\middle|\Psi^{(N)}_{\mathbf{n}}\right\rangle\left\langle \mathbf{n}^{(N)}\middle|\mathrm{BCS}\right\rangle}{\left\langle \mathrm{CBCS}\middle|\mathrm{CBCS}\right\rangle}\,. \tag{17}$$

It is helpful in the development of the formal theory to utilize methods developed for the cluster expansions of the normal system [15]. Besides the normalization integrals $I_{\mathbf{m},\mathbf{m}}$ defined in Eq. (7) we need the overlap integrals

$$M^{(N)}_{\mathbf{m},\mathbf{n}} \equiv \left\langle \Psi^{(N)}_{\mathbf{m}}\middle|\Psi^{(N)}_{\mathbf{n}}\right\rangle \equiv \delta_{\mathbf{m},\mathbf{n}} + N^{(N)}_{\mathbf{m},\mathbf{n}}\,. \tag{18}$$

and matrix elements of the Hamiltonian

$$H^{(N)}_{\mathbf{m},\mathbf{n}} \equiv \left\langle \Psi^{(N)}_{\mathbf{m}}\middle|\hat{H}'\middle|\Psi^{(N)}_{\mathbf{m}}\right\rangle \tag{19}$$

$$W^{(N)}_{\mathbf{m},\mathbf{n}} \equiv H^{(N)}_{\mathbf{m},\mathbf{n}} - \frac{1}{2}\left(H^{(N)}_{\mathbf{m},\mathbf{m}} + H^{(N)}_{\mathbf{n},\mathbf{n}}\right)M^{(N)}_{\mathbf{m},\mathbf{n}}\,. \tag{20}$$

Using these definitions, we can write the superfluid free energy as

$$\begin{aligned}\left\langle H'\right\rangle_s &= E_{\mathrm{diag}} + \frac{\sum_{N,\mathbf{m},\mathbf{n}}\left\langle \mathrm{BCS}\middle|\mathbf{m}^{(N)}\right\rangle W^{(N)}_{\mathbf{m},\mathbf{n}}\left\langle \mathbf{n}^{(N)}\middle|\mathrm{BCS}\right\rangle}{\left\langle \mathrm{CBCS}\middle|\mathrm{CBCS}\right\rangle}\\ &+ \frac{1}{2}\frac{\sum_{N,\mathbf{m}\neq\mathbf{n}}\left\langle \mathrm{BCS}\middle|\mathbf{m}^{(N)}\right\rangle\left(H^{(N)}_{\mathbf{m},\mathbf{m}} + H^{(N)}_{\mathbf{n},\mathbf{n}} - 2E_{\mathrm{diag}}\right)N_{\mathbf{m},\mathbf{n}}\left\langle \mathbf{n}^{(N)}\middle|\mathrm{BCS}\right\rangle}{\left\langle \mathrm{CBCS}\middle|\mathrm{CBCS}\right\rangle}\\ &\equiv E_{\mathrm{diag}} + E_{\mathrm{offd}} + E_{\mathrm{enum}}\end{aligned} \tag{21}$$

where

$$E_{\mathrm{diag}} = \sum_{N,\mathbf{m}}\left\langle \mathrm{BCS}\middle|\mathbf{m}^{(N)}\right\rangle H^{(N)}_{\mathbf{m},\mathbf{m}}\left\langle \mathbf{m}^{(N)}\middle|\mathrm{BCS}\right\rangle\,. \tag{22}$$

The actual expansions are tedious matter [22] and will be reported elsewhere, we display here only the most important results and discuss, in a simplified example, the salient conclusions of our analysis. The *physical interpretation* of the three terms in Eq. (21) is revealed by examining cluster expansions for the individual quantities. It turns out that the term E_{diag} is obtained simply the expectation value of a state where the momentum distribution of the ground state given by $v_{\mathbf{k}}^2$ instead of the normal Fermi distribution $v_{\mathbf{k}}^{(0)} = \theta(k_{\mathrm{F}} - k)$. The second term in Eq. (21) contains only off–diagonal matrix elements, we can order this term according to the *number of orbitals* in which the states $\left|\mathbf{m}^{(N)}\right\rangle$ and $\left|\mathbf{n}^{(N)}\right\rangle$ differ, which will turn out, in the diagrammatic analysis, the number of Cooper pairs. If pairing is weak as examined in our previous work, then $\left|\mathbf{m}^{(N)}\right\rangle$ and $\left|\mathbf{n}^{(N)}\right\rangle$ differ by only two states. In other words we consider the interaction of only one Cooper pair at a time.

The last term, E_{enum} gives rise to the "energy numerator corrections" shown in (11). A discussion of the significance of these terms is found in Refs. 9 and 4.

The purpose of the decomposition (21) is that cluster expansion and resummation techniques of $(E_{\mathrm{diag}} + E_{\mathrm{offd}})$ can be derived directly from the corresponding expression for the normal system. We cite here only the final result:

In the *normal* system, the cluster expansion of the generating function $G \equiv \ln I_{\mathbf{o},\mathbf{o}}$ is diagrammatically expressed by the sum of all irreducible diagrams constructed by the following rules [23]:

IOP Conf. Series: Journal of Physics: Conf. Series **1041** (2018) 012010 doi:10.1088/1742-6596/1041/1/012010

(i) Small circles depict particle coordinates. Filled circles imply integration over the particle coordinate and a density factor.
(ii) Every point is attached to the diagram by at least one correlation line $h(r) \equiv \exp(u_2(r))-1$.
(iii) Every pair of points is connected by at most one correlon line.
(iv) Exchange lines $\ell(rk_{\rm F})$ always appear in non-overlapping closed loops, or polygons. An $n-$sided exchange polygon contributes a weight factor $(-1/\nu)^{n-1}$ to the corresponding analytic expression, where ν is the degree of degeneracy of single–particle states.
(v) No point can be attached to more than two exchange lines.

The corresponding expansion for the *superfluid* system is generated from that by the following rule:

(i) Intepret the density factor as

$$\rho = \frac{\nu}{\Omega}\sum_{\mathbf{k}} v_{\mathbf{k}}^2\,, \tag{23}$$

where Ω is the normalization volume.

(ii) re-interpret all exchange lines $\ell(rk_{\rm F})$ as

$$\ell_v(r) \equiv \frac{\nu}{\rho\Omega}\sum_{\mathbf{k}} v_{\mathbf{k}}^2 e^{{\rm i}\mathbf{k}\cdot\mathbf{r}} = \frac{\nu}{\rho}\int\frac{d^3k}{(2\pi)^3} v_{\mathbf{k}}^2 e^{{\rm i}\mathbf{k}\cdot\mathbf{r}}\,. \tag{24}$$

(iii) In each exchange loop, replace, in turn, each pair of exchange lines $\ell_v(r_{ij})\ell_v(r_{kl})$ by a pair $-\ell_u(r_{ij})\ell_u(r_{kl})$, where

$$\ell_u(r) \equiv \frac{\nu}{\rho\Omega}\sum_{\mathbf{k}} u_{\mathbf{k}} v_{\mathbf{k}} e^{{\rm i}\mathbf{k}\cdot\mathbf{r}} = \frac{\nu}{\rho}\int\frac{d^3k}{(2\pi)^3} u_{\mathbf{k}} v_{\mathbf{k}} e^{{\rm i}\mathbf{k}\cdot\mathbf{r}}\,. \tag{25}$$

The FHNC theory for the terms $E_{\rm diag} + E_{\rm offd}$ in the energy expression (21) can therefore be read off from the diagrammatic expansions for the normal system. As an example, we show in Fig. 3 the diagrammatic representation of some leading potential energy contributions to $E_{\rm diag} + E_{\rm offd}$.

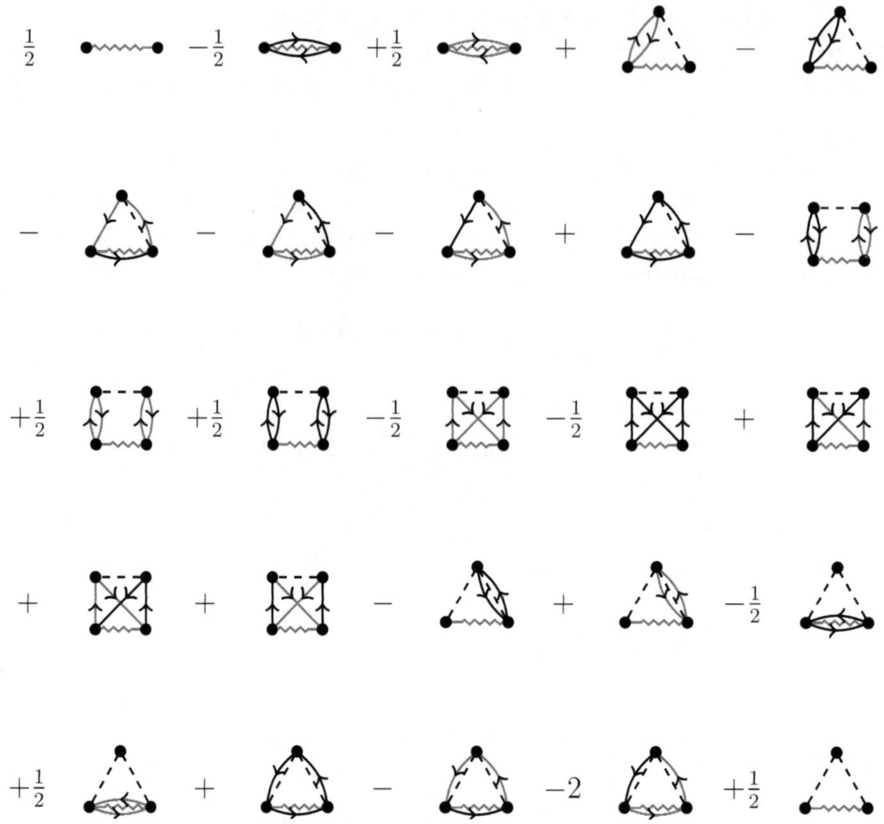

Figure 3. (color online) The figure shows the leading diagrams contributing to the potential energy. The black and red exchange lines depict the exchange functions $\ell_v(r_{ij})$ and $\ell_u(r_{ij})$, *cf.* Eqs. (24) and (25), respectively, and the red wavy line an interaction $V(r_{ij})$. Otherwise we follow the usual diagrammatic conventions [23]. The uniform limit approximation amounts to keeping only diagrams 1-5, 10-12, and 25. The term $E_{\rm diag}$ is represented by the subset of diagrams containing only black exhange lines.

The energy numerator terms must be treated separately, we will spell out in the next section the simplest possible form.

2.1. Uniform limit approximation

We shall not implement the full method here but rather spell out what is known as the "uniform limit" approximation because it shows already the most important features of the theory and implications of using locally correlated wave functions in the superfluid system. Technically, the uniform limit approximation amounts to assuming that a product of any two correlation functions is negligible when it appears in coordinate space, but not negligible in momentum space, in other words we assume $h^2(r) \ll h(r)$ but not that $\tilde{h}^2(k)$ is small compared to $\tilde{h}(k)$. Note that we define here the dimensionless Fourier transform

$$\tilde{f}(k) = \rho \int d^3r e^{\mathrm{i}\mathbf{r}\cdot\mathbf{k}} f(r)\,. \tag{26}$$

In the uniform limit approximation, only those exchange diagrams are retained that contain either a loop $\ell_v^2(r_{ij})$ or $\ell_u^2(r_{ij})$, see Fig. 3. It is then convenient to define

$$\sigma_v(r) = \frac{1}{\rho}\delta(r) - \frac{1}{\nu}\ell_v^2(r) \tag{27}$$

$$\sigma_u(r) = \frac{1}{\nu}\ell_u^2(r) \,. \tag{28}$$

Then, the static structure function of the "non–interactig" system is

$$S_{\rm F}(k) = \frac{1}{\langle N\rangle_0}\langle {\rm BCS}|\rho_{\mathbf{k}}\rho_{-\mathbf{k}}|{\rm BCS}\rangle = \tilde\sigma_u(k) + \tilde\sigma_v(k) \tag{29}$$

$$= 1 - \frac{1}{\nu}\left[\ell_v^2(r)\right]^{\mathcal F}(k) + \frac{1}{\nu}\left[\ell_u^2(r)\right]^{\mathcal F}(k) \tag{30}$$

where $[\ldots]^{\mathcal F}(k)$ is used as an alternative notation for the Fourier–transform (26) and $\langle N\rangle_0 = \nu\sum_{\mathbf{k}} v_{\mathbf{k}}^2$ is the expectation value of the number operator $\hat N$ with respect to the BCS state (5).

$$\frac{E_{\rm diag} + E_{\rm offd}}{N} = \frac{T_{\rm BCS}}{N} + \frac{1}{2}\tilde V(0+) + \frac{1}{2}\int\frac{d^3k}{(2\pi)^3\rho}\left[S(k) - 1\right]\tilde V(k) + \frac{1}{2}\int\frac{d^3k}{(2\pi)^3\rho}\frac{t(k)}{2}\frac{(S(k) - S_{\rm F}(k))^2}{S_{\rm F}^2(k)S(k)}\,, \tag{31}$$

where $t(k) = \frac{\hbar^2k^2}{2m}$ and $T_{\rm BCS} = \nu\sum_{\mathbf{k}} t(k)v_{\mathbf{k}}^2$ is the kinetic energy associated with the BCS state (5). In this approximation, the static structure function is given by the chain approximation,

$$S(k) = \frac{S_{\rm F}(k)}{1 - \tilde h(k)S_{\rm F}(k)}\,. \tag{32}$$

We can, in this case, simply copy the equations from previous work, with the understanding that the static structure function of the non–interacting system, $S_{\rm F}(k)$, is replaced by (30).

The evaluation of the energy numerator expression $E_{\rm enum}$ must be done in an independent cluster expansion. These terms lead, in the weakly coupled limit considered in previous work, to the second term in the pairing matrix element (11). Recall, however, the derivation of the pairing matrix elements in Ref. 15 and the analysis of Ref. 9 that the matrix elements $\mathcal{W}_{\mathbf{k},\mathbf{k}'}$ should be identified with a static approximation of the 2-body scattering matrix which contains ring– and ladder–diagrams. In the uniform limit approximation, the pairing interaction, on the other hand, does not contain ladder diagrams, and the energy numerator term cancels the overcounted ladder diagrams in the scattering matrix. The analysis is analogous to the one found in Ref. 24 where the gap equation is expressed in terms of the T-matrix.

Now the "uniform limit" approximation contains only chain diagrams, hence there is no double-counting of ladder diagrams and the energy numerator term should, in that approximation, not be retained for the derivation of the Euler equation. We have therefore now two optimization conditions, one for the Bogoliubov amplitudes and one for the pair correlations. The gap equation (14) remains the same, only the pairing matrix element depends implicitly on the Bogoliubov amplitudes. The analysis of the Euler equation for the spatial amplitudes is more complicated and will be carried out in the next section.

3. Euler equation for local correlations

The analysis of the Euler equation for the spatial correlations turns out to be much more subtle: Carrying out the variation of the energy (31) yields the familiar answer

$$S(k) = \frac{S_{\rm F}(k)}{\sqrt{1+\frac{2S_{\rm F}^2(k)}{t(k)}\tilde V(k)}}\,. \tag{33}$$

(Retaining the energy numerator terms simply supplements the bare interaction with a small additive term). At the first glance, the result looks innocuous. In the *normal* system, we have $S_{\rm F}(k) \propto k$ as $k \to 0+$ and the condition that the expression under the square root be positive

$$1 + \frac{3}{4}\frac{3m}{\hbar^2 k_{\rm F}^2}\tilde V(0+) > 0. \tag{34}$$

which is — apart from the numerical factor 3/4 which will be discussed momentarily — recognized as the condition $F_0^s > -1$. In the *superflud* system, we have instead

$$S_{\rm F}(0+) = 2\frac{\sum_{\bf k} u_{\bf k}^2 v_{\bf k}^2}{\sum_{\bf k} v_{\bf k}^2} > 0\,. \tag{35}$$

This says that a locally correlated wave function for a superfluid system has a spurious instability whenever $\tilde V(0+) < 0$, no matter how small the gap is. This is evidently only a consequence of the approximations implicit to the wave function (6). We have shown this here only for the case of a weakly interacting system, but it is also true in the general case that all FHNC diagrams, and possibly also higher–order correlation functions, are included: The Euler equation remains structurally the same, only the bare potential is replaced by an effective interaction [2]. Our observation applies, of course, equally to “fixed–node” Monte Carlo calculations which may see this instability only in large stochastic fluctuations. In view of the fact that the BCS-BEC crossover is driven by a net attractive interaction (see also the right panel of Fig. 1), it is worth examining this problem in detail.

The source of the problem can be identified as follows: In the *normal* system, the expression (33) can be derived from an ordinary random phase approximation

$$S(k) = -\int_0^\infty \frac{d\omega}{\pi}\Im m\frac{\chi_0(k,\omega)}{1-\tilde V(q)\chi_0(k,\omega)} \tag{36}$$

where $\chi_0(k,\omega)$ is the Lindhard function. Consistent with the convention (26) according to which $\tilde V(q)$ has the dimension of an energy, we have defined the density-density response function slightly different than usual [16] namely such that has the dimension of an inverse energy. Eq. (33) can be obtained by approximating the Lindhard function $\chi_0(q,\omega)$ by a “collective” Lindhard function (occasionally also referred to as “one–pole approximation” or “mean spherical approximation”) $\chi_0^{\rm coll}(q,\omega)$ which is constructed by approximating the particle–hole band by an effective single pole such that the m_0 and m_1 sum rules are satisfied [2]

$$-\Im m\int_0^\infty \frac{d\omega}{\pi}\chi_0^{\rm coll}(k,\omega) = -\Im m\int_0^\infty \frac{d\omega}{\pi}\chi_0(k,\omega) = S_{\rm F}(k) \tag{37}$$

$$-\Im m\int_0^\infty \frac{d\omega}{\pi}\omega\chi_0^{\rm coll}(k,\omega) = -\Im m\int_0^\infty \frac{d\omega}{\pi}\omega\chi_0(k,\omega) = t(k)\,. \tag{38}$$

This leads to

$$\chi_0^{\rm coll}(k,\omega) = \frac{2t(k)}{(\omega+{\rm i}\eta)^2 - \left(\frac{t(k)}{S_{\rm F}(k)}\right)^2}\,. \tag{39}$$

The frequency integration (36) then can be carried out analytically and leads to the equation (33). Experience with electronic systems and the very strongly interaction ^{3}He has shown that the expression (33) is accurate within 1-2 percent; this is the reason why the Jastrow–Feenberg theory for fermions has been so successful. Of course, this is relevant only for integrated or Fermi–sea averaged quantities; we have pointed out already a long time ago [25] that the approximation should be particularly poor for observables that are determined by the dynamics close to the Fermi surface. Among others, the stability condition (34) is replaced by

$$1 + \frac{3m}{\hbar^2 k_{\rm F}^2}\tilde V(0+) > 0. \tag{40}$$

which is the correct stability condition.

Returning to the superfluid case, one can expect that the correct Lindhard function removes the spurious instability. There have been several suggestions [26, 27, 28, 29]. The most frequently used form for $T = 0$ is, in terms of the usual relationships of BCS theory,

$$\begin{aligned} u_k^2 &= \frac{1}{2}\left(1+\frac{\xi_{\mathbf k}}{E_{\mathbf k}}\right)\\ v_k^2 &= \frac{1}{2}\left(1-\frac{\xi_{\mathbf k}}{E_{\mathbf k}}\right)\,. \end{aligned} \tag{41}$$

with $\xi_{\mathbf k} = t(k) - \mu$ and $E_{\mathbf k} = \sqrt{\xi_{\mathbf k}^2 + \Delta_{\mathbf k}^2}$ we have [30, 31, 32, 26]

$$\chi_0(\mathbf q,\omega) = \frac{\nu}{\langle N\rangle_0}\sum_{\mathbf k} b^-_{\mathbf k,\mathbf q}\left[\frac{1}{\omega - E_{\mathbf k+\mathbf q} - E_{\mathbf k} + {\rm i}\eta} - \frac{1}{\omega + E_{\mathbf k+\mathbf q} + E_{\mathbf k} + {\rm i}\eta}\right] \tag{42}$$

where

$$b^-_{\mathbf k,\mathbf q} = \frac{1}{4}\left[1 - \frac{\xi_{\mathbf k}\,\xi_{\mathbf k+\mathbf q}}{E_{\mathbf k}\,E_{\mathbf k+\mathbf q}} + \frac{\Delta_{\mathbf k}\,\Delta_{\mathbf k+\mathbf q}}{E_{\mathbf k}\,E_{\mathbf k+\mathbf q}}\right]\,. \tag{43}$$

In the limit of a normal system, the coefficients $b^-_{\mathbf k,\mathbf q}$ become

$$b^-_{\mathbf k,\mathbf q} \to n_{\mathbf k}(1 - n_{\mathbf k+\mathbf q})\,, \tag{44}$$

where $n_{\mathbf k}$ is the Fermi distribution, as it should come out.

This superfluid Lindhard function is consistent with the $S_{\rm F}(k)$ as defined in Eq. (29).

$$\begin{aligned} S_{\rm F}(k) &= -\int_0^\infty \frac{d\omega}{\pi}\Im m\chi_0(k,\omega) = \frac{\nu}{\langle N\rangle_0}\int_0^\infty d\omega\sum_{\mathbf q} b^-_{\mathbf q,\mathbf k}\delta(\omega - E_{\mathbf q} - E_{\mathbf q+\mathbf k})\\ &= \frac{\nu}{\langle N\rangle_0}\sum_{\mathbf q} b^-_{\mathbf q,\mathbf k} = \tilde\sigma_u(k) + \tilde\sigma_v(k) \end{aligned} \tag{45}$$

We can now return to the frequency integration (36). All we need to show is that this expression exists, for small gaps, for $-1 < \frac{3m}{\hbar^2 k_{\rm F}^2}\tilde V(0+)$. For that, it is sufficient to look at the limit $q \to 0$ of the *static* response function and the *static* Lindhard function.

IOP Conf. Series: Journal of Physics: Conf. Series **1041** (2018) 012010 doi:10.1088/1742-6596/1041/1/012010

For $\omega = 0$ we get for the static Lindhard function

$$\lim_{q\to 0}\chi_0(q,0) = -\frac{\nu}{2\rho}\int\frac{d^3k}{(2\pi)^3}\frac{\Delta_{\mathbf{k}}^2}{E_{\mathbf{k}}^3}\,. \tag{46}$$

To estimate this limit for small gap energies, we add and subtract a function that can be integrated exactly such that the remainder vanishes as $\Delta \to 0$.

$$\begin{aligned}
\lim_{q\to 0}\chi_0(q,0) &= -\frac{4\pi\nu}{2\rho}\int\frac{dkk^2}{(2\pi)^3}\frac{\Delta_{\mathbf{k}}^2}{\left((t(k)-\mu)^2+\Delta_{\mathbf{k}}^2\right)^{3/2}}\\
&= -\frac{3}{2k_{\rm F}^3}\int dk\left[\frac{k^2\Delta_{\mathbf{k}}^2}{\left((t(k)-\mu)^2+\Delta_{\mathbf{k}}^2\right)^{3/2}}-\frac{kk_{\rm F}\Delta^2}{\left((t(k)-\mu)^2+\Delta^2\right)^{3/2}}\right.\\
&\qquad\left.+\frac{kk_{\rm F}\Delta^2}{\left((t(k)-\mu)^2+\Delta^2\right)^{3/2}}\right]\\
&= \mathcal{O}(\Delta)-\frac{3}{4}\left[\frac{1}{\sqrt{\mu^2+\Delta^2}}+\frac{1}{\mu}\right]\to-\frac{3}{2\mu}\quad\text{as}\quad\Delta\to 0\,.
\end{aligned}\tag{47}$$

This is identical to the same limit of the Lindhard function of the normal system, see Fetter-Walecka Eq. (12.46b), considering that the Lindhard function defined here contains a factor $1/\rho$.

With that we get for the stability condition

$$1+\frac{3\tilde{V}(0+)}{2\mu} > 0 \tag{48}$$

4. Discussion

We have formulated in this contribution the beginnings of a theoretical method to deal with strongly interacting superfluids that promises the same accuracy as what was achieved for the normal helium liquids. We have studied the Euler equation for the local correlations in a simplified example that does not require the formulation of the full FHNC theory. We have derived the interesting, yet disturbing, result that the Euler equation has no physically meaningful solution for net attractive interactions $\tilde{V}(0+) < 0$. We hasten to point out that this result is not specific to the "uniform limit approximation", in the fully correlated theory the bare interaction $\tilde{V}(q)$ is simply replaced by the particle–hole interaction. This is a rather profound observation since it applies to any local correlation operator F; including the "fixed–node" approximation used in Quantum Monte Carlo calculations where it would, of course, be hard to discover.

We have also outlined a pathway to the solution of this problem: One must go beyond local correlations. The *rigorous* way to do that has been carried out in Ref. 33: One must add, to the variational energy expectation value, an infinite sum of terms in CBF perturbation theory. That has the effect that the "collective approximation" for $S(k)$, Eq. (33) is replaced by the RPA expression (36). The calculations of Ref. 33 have been rather tedious; to be rigorous one would have to re-do all of this work for the superfliud system. The formulation of Coupled Cluster theory with correlated wave functions [6] would be a way to carry this through, but, on the other hand, our result is sufficiently plausible to be applicable without a rigorous proof.

Acknowledgments

This work was supported, in part, by the College of Arts and Sciences, University at Buffalo SUNY, and the Austrian Science Fund project I602 (to EK). We would like to thank John Clark and Peter Schuck for useful discussions on the subject of this paper.

References

[1] Feenberg E 1969 *Theory of Quantum Fluids* (New York: Academic)

[2] Krotscheck E 2000 *J. Low Temp. Phys.* **119** 103–145

[3] Egger J, Krotscheck E and Zillich R E 2011 *J. Low Temp. Phys.* **165** 275–291

[4] Fan H H, Krotscheck E and Clark J W 2017 *J. Low Temp. Phys.* **189** 470–494

[5] Krotscheck E and Clark J W 1980 *Nucl. Phys. A* **333** 77–115

[6] Krotscheck E, Smith R A and Jackson A D 1981 *Phys. Rev. B* **24** 6404–6420

[7] Duine R and Stoof H T C 2004 *Phys. Rep.* **396** 115–195

[8] Chen Q, Stajic J, Tan S and Levin K 2005 *Phys. Rep.* **412** 1–88

[9] Fan H H, Krotscheck E, Lichtenegger T, Mateo D and Zillich R E 2015 *Phys. Rev. A* **92** 023640

[10] Reid, Jr R V 1968 *Ann. Phys. (NY)* **50** 411–448

[11] Wiringa R B, Stoks V G J and Schiavilla R 1995 *Phys. Rev. C* **51** 38–51

[12] Jackson A D, Lande A and Smith R A 1982 *Phys. Rep.* **86** 55–111

[13] Jackson A D, Lande A and Smith R A 1985 *Phys. Rev. Lett.* **54** 1469–1471

[14] Krotscheck E, Smith R A and Jackson A D 1986 *Phys. Rev. A* **33** 3535–3536

[15] Krotscheck E and Clark J W 1979 *Nucl. Phys. A* **328** 73–103

[16] Fetter A L and Walecka J D 1971 *Quantum Theory of Many-Particle Systems* (New York: McGraw-Hill)

[17] Cooper L N, Mills R L and Sessler A M 1959 *Phys. Rev.* **114** 1377–1382

[18] Noziéres P and Schmitt-Rink S 1985 *J. Low Temp. Phys.* **59** 195–211

[19] Gezerlis A and Carlson J 2008 *Phys. Rev. C* **77** 032801

[20] Gezerlis A and Carlson J 2010 *Phys. Rev. C* **81** 025803

[21] Krotscheck E 2002 *Introduction to Modern Methods of Quantum Many–Body Theory and their Applications* (*Advances in Quantum Many–Body Theory* vol 7) ed Fabrocini A, Fantoni S and Krotscheck E (Singapore: World Scientific) pp 267–330

[22] Fan H H 2018 *Pairing Phenomena from Low-Density Fermi Gases to Neutron Star Matter* Ph.D. thesis University at Buffalo SUNY

[23] Clark J W 1979 *Progress in Particle and Nuclear Physics* vol 2 ed Wilkinson D H (Oxford: Pergamon Press Ltd.) pp 89–199

[24] Pethick C J and Smith H 2008 *Bose-Einstein Condensation in Dilute Gases* second edition ed (Cambridge, UK: Cambridge University Press)

[25] Jackson A D, Krotscheck E, Meltzer D and Smith R A 1982 *Nucl. Phys. A* **386** 125–165

[26] Voo K K, Wu W C, Li J X and Lee T K 2000 *Phys. Rev. B* **13** 9095–9100

[27] Combescot R, Kagan M Y and Stringari S 2006 *Phys. Rev. A* **74** 042717

[28] Steiner A W and Reddy S 2009 *Phys. Rev. C* **79** 015802

[29] Vitali E, Shi H, Qin M and Zhang S 2017 *J. Low Temp. Phys.* **189** 312–327

[30] Schrieffer J R 1999 *Theory Of Superconductivity (Advanced Books Classics)* revised ed (Perseus Books)

[31] Kee H Y and Varma C M 1998 *Phys. Rev. B* **58** 15035–15044

[32] Kee H Y and Kim Y B 1999 *Phys. Rev. B* **59** 4470–4474

[33] Krotscheck E 1982 *Phys. Rev. A* **26** 3536–3556

IOP Conf. Series: Journal of Physics: Conf. Series **1041** (2018) 012011 doi:10.1088/1742-6596/1041/1/012011

2D Dipolar Bosons with Non-zero Tilt Angles

Pengtao Shen and Khandker Quader

Department of Physics, Kent State University, Kent, OH 44242, USA

E-mail: quader@kent.edu

Abstract. Recent experimental advances in creating stable dipolar systems, including polar molecules with large electric dipole moments, have led to vigorous theoretical activities. Here we discuss our work on homogeneous 2D dipolar bosons with dipoles oriented at an angle to the direction perpendicular to the confining 2D plane, i.e. for non-zero tilt angles. Using Bogoliubov-de Gennes equations, we obtain T=0 excitation spectrum of the dipolar Bose gas, and explore instabilities at varying tilt angle, density and dipolar coupling. We map out a phase diagram a function of tilt angle, dipole strength and density. We find the development of maxon-roton behavior leading to roton instabilities at large densities for small tilt angles, and at low densities for large tilt angles. The behavior is anisotropic in k-space; accordingly the roton instabilities occur in the k_y direction, suggestive of inhomogeneity and stripe phase, with density mode becoming soft in the y direction. Our results compare favorably with existing Quantum Monte Carlo calculations for non-zero tilt angles.

1. Introduction

In recent years, systems with long-range and anisotropic interaction have generated considerable interest. Among these are dipolar interactions between Bosonic atoms or polar molecules. Examples are ^{52}Cr [1], $^{87}Rb^{133}Cs$ [2], $^{41}K^{87}Pb$, among which RbCS in particular has sizable electric dipole moment $\sim$ 1.28 Debye. In general, the electric dipole moments of the polar molecules are substantially larger than the magnetic dipole moments of atoms. The nature of the dipolar interaction can give rise to novel quantum phases. Experimentally, low dimensional cold gas system has been realized by optical trapping. A recent reporting [3] of observation of roton mode in dipolar system provides added motivation for theoretical exploration. Since applied electric field can be used to fix the orientation of 2D dipoles in a system, one of the degrees of freedom in these systems is the tilt angle fixed by the direction of the applied electric field. Much of the work on quasi 2D and other 2D Bose systems has been done for zero tilt angle [4, 5], except for Quantum Monte Carlo (QMC) calculations [6, 7] on 2D bosons.

While it is possible to realize various kinds of low dimensional systems, here we focus on homogeneous 2D Bose gas at zero temperature, for arbitrary tilt angles. Among our objectives is to test how well mean-field calculations such Hartree-Fock-Bogoliubov (HFB) compare with the findings of QMC calculations. As a background for our work, we first review the Hartree Fock Bogoliubov (HFB) theory in the next section.

2. Background: Hartree Fock Bogoliubov (HFB) theory

The grand canonical Hamiltonian of the interacting Bose gas is given by:

$$\begin{aligned}\hat{K} \equiv \hat{H} - \mu\hat{N} &= \int dr\hat{\psi}^{\dagger}(r)[-\frac{\nabla^2}{2m} + U_{ex}(r) - \mu]\hat{\psi}(r) \\ &+ \frac{1}{2}\int drdr'\hat{\psi}^{\dagger}(r)\hat{\psi}^{\dagger}(r')V(r-r')\hat{\psi}(r)\hat{\psi}(r') \end{aligned} \tag{1}$$

where $\hat{\psi}(r)$ is the annihilation Bose field operator. The field operator is expanded as: $\hat{\psi}(r) = \phi_0(r) + \tilde{\psi}(r)$, with $\phi_0(r) = \langle\hat{\psi}(r)\rangle$ being the condensate wave function and $\tilde{\psi}(r) = \hat{\psi}(r) - \phi_0(r)$ the fluctuation from the condensate. Substituting the expansion into Eq.(1), the grand canonical Hamiltonian $\hat{K}$ can be expanded in orders of $\tilde{\psi}$.

In Bogoliubov approximation, the expansion is taken upto second order of $\tilde{\psi}$. $K = K_0 + K_1 + K_2$, where

$$\hat{K}_0 = \int dr\phi_0^*(r)[-\frac{\nabla^2}{2m} + U_{ex}(r) - \mu]\phi_0(r) \tag{2}$$

$$\begin{aligned}\hat{K}_1 &= \int dr\phi_0^*(r)[-\frac{\nabla^2}{2m} + U_{ex}(r) - \mu]\tilde{\psi}(r) \\ &+ \int dr\tilde{\psi}^{\dagger}(r)[-\frac{\nabla^2}{2m} + U_{ex}(r) - \mu]\phi_0(r) \\ &+ \int drdr'|\phi_0(r')|^2\phi_0^*(r)V(r-r')\tilde{\psi}(r) \\ &+ \int drdr'|\phi_0(r')|^2\tilde{\psi}^{\dagger}(r)V(r-r')\phi_0(r) \end{aligned} \tag{3}$$

$$\begin{aligned}\hat{K}_2 &= \int dr\tilde{\psi}^{\dagger}(r)[-\frac{\nabla^2}{2m} + U_{ex}(r) - \mu]\tilde{\psi}(r) \\ &+ \int drdr'|\phi_0(r')|^2\tilde{\psi}^{\dagger}(r)V(r-r')\tilde{\psi}(r) \\ &+ \int drdr'\phi_0^*(r)\phi_0(r')\tilde{\psi}^{\dagger}(r')V(r-r')\tilde{\psi}(r) \\ &+ \frac{1}{2}\int drdr'\tilde{\psi}^{\dagger}(r)\tilde{\psi}^{\dagger}(r')V(r-r')\phi_0(r)\phi_0(r') \\ &+ \frac{1}{2}\int drdr'\phi_0^*(r)\phi_0^*(r')V(r-r')\tilde{\psi}(r)\tilde{\psi}(r') \end{aligned} \tag{4}$$

K_1 has to vanish so as to minimize the grand canonical Hamiltonian, so one obtains the equation

$$[-\frac{\nabla^2}{2m} + U_{ex}(r) + \int dr'|\phi_0(r')|^2 V(r-r')]\phi_0(r) = \mu\phi_0(r) \tag{5}$$

This is the generalized Gross-Pitaevskii equation, which gives the condensate wave function $\phi_0(r)$.

IOP Conf. Series: Journal of Physics: Conf. Series **1041** (2018) 012011 doi:10.1088/1742-6596/1041/1/012011

Combining K_0 and K_2, and defining $T(r) = -\frac{\nabla^2}{2m} + U_{ex}(r) - \mu$, one obtains the Hartree-Fock-Bogoliubov (HFB) Hamiltonian:

$$\begin{aligned}
\hat{K}_{HFB} &= \int dr\phi_0^*(r)T(r)\phi_0(r) \\
&+ \int dr\tilde{\psi}^\dagger(r)T(r)\tilde{\psi}(r) \\
&+ \int drdr'|\phi_0(r')|^2\tilde{\psi}^\dagger(r)V(r-r')\tilde{\psi}(r) \\
&+ \int drdr'\phi_0^*(r)\phi_0(r')\tilde{\psi}^\dagger(r')V(r-r')\tilde{\psi}(r) \\
&+ \frac{1}{2}\int drdr'\tilde{\psi}^\dagger(r)\tilde{\psi}^\dagger(r')V(r-r')\phi_0(r)\phi_0(r') \\
&+ \frac{1}{2}\int drdr'\phi_0^*(r)\phi_0^*(r')V(r-r')\tilde{\psi}(r)\tilde{\psi}(r')
\end{aligned} \tag{6}$$

Bogoliuvbov transformation $\tilde{\psi}(r) = \sum[u_i(r)\hat{a}_i - v_i^*(r)\hat{a}_i^\dagger]$ diagonalizes the Hamiltonian, with $u_i(r)$ and $v_i(r)$ satisfying the equation

$$\begin{aligned}
\varepsilon_j u_j(r) &= (\hat{T}(r) + \int dr'|\phi_0(r')|^2 V(r-r'))u_j(r) \\
&+ \int dr'V(r-r')\phi_0^*(r)\phi_0(r')u_j(r') \\
&+ \int dr'\phi_0(r)\phi_0(r')V(r-r')v_j(r')
\end{aligned} \tag{7}$$

$$\begin{aligned}
-\varepsilon_j v_j(r) &= (\hat{T}(r) + \int dr'|\phi_0(r')|^2 V(r-r'))v_j(r) \\
&+ \int dr'V(r-r')\phi_0^*(r)\phi_0(r')v_j(r') \\
&+ \int dr'\phi_0^*(r)\phi_0^*(r')V(r-r')u_j(r')
\end{aligned} \tag{8}$$

In homogenous case, $U_{ex}(r) = 0$; the ground state is $\phi_0(r) = \phi_0$; the condensate density is $\int dr\phi_0^2(r) = n_0$; and $\mu = \int dr'|\phi_0(r')|^2 V(r-r')] = n_0 V(q)$. The Bogoliubov transform of the fluctuation is $\tilde{\psi}(r) = \sum[e^{-iqr}\hat{a}_q - e^{iqr}\hat{a}_q^\dagger]$ Substituting $\tilde{\psi}(r)$ into Eq.(7) and Eq.(8), and Fourier transforming $V(r-r')$, one obtains

$$\begin{aligned}
\varepsilon_q e^{-iqr} &= (q^2/2m + V(q)n_0)e^{-iqr} + V(q)n_0 e^{iqr} \\
-\varepsilon_q e^{iqr} &= (q^2/2m + V(q)n_0)e^{iqr} + V(q)n_0 e^{-iqr}
\end{aligned} \tag{9}$$

This gives the Bogoliubov spectrum

$$\varepsilon(q) = \sqrt{(q^2/2m)^2 + n_0 V(q)q^2/m} \tag{10}$$

An equivalent result can be derived in the random phase approximation (RPA) approach. If in the Hartree approximation the density-density response is $\chi_0(k,w)$, then the RPA response function is given by:

$$\chi_{RPA}(k,w) = \frac{\chi_0(k,w)}{1 - V_k\chi_0(k,w)} \tag{11}$$

For a free boson system with condensation, and $n_0 \approx n$,

$$\chi_0(k,w) = \frac{n_0}{\hbar w - \epsilon_k} - \frac{n_0}{\hbar w + \epsilon_k} \tag{12}$$

where n_0 is the condensate density, and $\epsilon_k = \hbar^2 k^2/2m$ is the free particle kinetic energy. Substitute Eq.(12) into Eq.(11), one can obtain the density-density response function of the interacting boson system with a condensate,

$$\chi_0(k,w) = \frac{2n_0\epsilon_k}{\hbar w^2 - \epsilon_k^2 - 2n_0 V_k \epsilon_k} \tag{13}$$

The pole of the above equation, $\hbar w = \sqrt{\epsilon_k^2 + 2n_0 V_k \epsilon_k}$, is exactly the Bogoliubov spectrum obtained we in Eq.(10).

3. Dipolar Bosons in 2 dimension

Consider a homogenous system of dipolar bosons in XY plane with tilt angle θ which is confined by a harmonic trap in z direction. The projection of the dipole is in x direction.

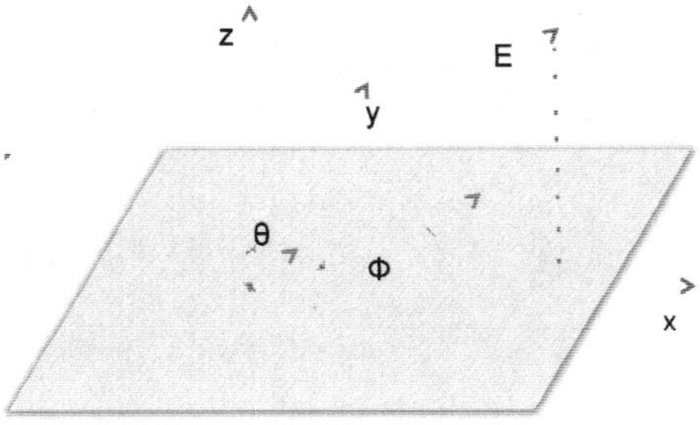

Figure 1. 2d dipole in x-y plane with tilt angle θ that defines the direction of electric field E relative to the z-direction. Φ is the angle in x-y plane, relative to x-direction.

The 2D dipole-dipole interaction is:

$$\begin{aligned} V_{2d}(r) &= \frac{d^2}{r^3}(1 - 3\sin^2\theta\cos^2\phi) \\ &= \frac{d^2}{r^3}[P_2(\cos\theta) - \frac{3}{2}\sin^2\theta\cos 2\phi] \end{aligned} \tag{14}$$

where ϕ is the azimuthal angle relative to x direction, θ the tilt angle, and d strength of the dipolar interaction. The Fourier transform of 2D dipolar interaction in momentum space is:

$$V(q) = \int_0^\infty d^2 r V(r) e^{iqr} \tag{15}$$

The 2D integral however is divergent in the limit $r \to 0$. One way to regularize [8] this is to introduce a short-range cut-off, r_c in the Fourier transform. The choice of a value for r_c can be motivated in the case of polar molecules, at least, by taking r_c to be of the order of the size of the molecule. It typically would be of the order of several Bohr radius, a_0. Performing the integral $\int_{r_c}^\infty$, the 2D dipole-dipole interaction (DDI) in momentum space can be written as $V(q) = V_s + V_l(q)$, with

$$\begin{aligned} V_s &= 2\pi d^2 \frac{P_2(\cos\theta)}{r_c} \\ V_l(q) &= -2\pi d^2 q(\cos^2\theta - \sin^2\theta\cos^2\phi) \end{aligned} \tag{16}$$

The short range part of the interaction V_s depend on the tilt angle θ: V_s is positive at small θ and becomes negative at large θ, The long range term, $V_l(q)$ has linear q dependence. At zero tilt angle, $V_l(q)$ is isotropic and negative in all direction. For non-zero tilt angle, $V_l(q)$ become anisotropic. In x direction, $V_l(q)$ gets less negative and becomes positive when $\theta > \pi/4$. In y direction, it is always negative. As will be seen, interesting consequences of 2D DDI, viz. instabilities of the system, happen when the DDI gets attractive in some direction. Hence, it is most interesting to consider the behavior in the y direction, i.e. $\phi = \pi/2$, when the DDI is most attractive. Following standard practice, we define the dipolar interaction length $a_{dd} = \frac{\bar{h}d^2}{m}$. a_{dd} can vary from $10a_0$ for magnetic dipoles to $10^4 a_0$ for electric dipoles. On scaling with a_{dd}, we obtain

$$\begin{aligned} V(q,\phi) &= 2\pi d^2 \left[\frac{P_2(\cos\theta)}{r_c} - q(\cos^2\theta - \sin^2\theta\cos^2\phi)\right] \\ &= 2\pi\frac{\bar{h}}{m}\left[\frac{a_{dd}}{r_c}P_2(\cos\theta) - qa_{dd}(\cos^2\theta - \sin^2\theta\cos^2\phi)\right] \\ V(q,\phi=\pi/2) &= 2\pi d^2\left[\frac{P_2(\cos\theta)}{r_c} - q\cos^2\theta\right] \\ &= 2\pi\frac{\bar{h}}{m}\left[\frac{a_{dd}}{r_c}P_2(\cos\theta) - qa_{dd}\cos^2\theta\right] \end{aligned} \tag{17}$$

4. Spectrum and Stability

Using Eq.(10) in Eq.(16), we can immediately obtain the HFB spectrum. On scaling with the dipolar strength a_{dd}, $\tilde{\varepsilon} = \varepsilon/\frac{\bar{h}}{ma_{dd}^2}$, $\tilde{q} = qa_{dd}$ and $\tilde{n} = n_0 a_{dd}^2$, we obtain a spectrum with dimensionless energy and momentum:

$$\tilde{\varepsilon}(\tilde{q},\phi) = \sqrt{\frac{\tilde{q}^4}{4} + 2\pi\tilde{q}^2\tilde{n}\left[\frac{a_{dd}}{r_c}P_2(\cos\theta) - \tilde{q}(\cos^2\theta - \sin^2\theta\cos^2\phi)\right]} \tag{18}$$

The spectrum is anisotropic and spectrum energy in y direction is lower than any other direction.

$$\tilde{\varepsilon}(\tilde{q},\phi=\pi/2) = \sqrt{\frac{\tilde{q}^4}{4} + 2\pi\tilde{q}^2\tilde{n}\left[\frac{a_{dd}}{r_c}P_2(\cos\theta) - \tilde{q}\cos^2\theta\right]} \tag{19}$$

We can now study the stability of the 2D dipolar Bose gas by considering the above expression for the spectrum.

A. $\theta < \cos^{-1}\frac{1}{\sqrt{3}}$

For $\theta < \cos^{-1}\frac{1}{\sqrt{3}} = 0.955$, the short range interaction is positive and the long range part in y direction is negative. The spectrum is always real and positive at small momentum; see Fig. 2. However, at sufficiently large density, n, a roton-maxon characteristic, reminiscent of that in ^4He, develops in the spectrum. For even larger density, the roton spectra becomes imaginary in y direction; this can be seen as "holes" at symmetric $\pm q$ region in the 3D plot in Fig. 3; the "holes" signifying that the energy ϵ is no longer real. The corresponding 2D plot depicts this feature as negative energy for $\phi = \pi/2$, i.e. the y-direction. At the same time, the spectra in the x-direction remain real, positive. This means the existence of density waves along one direction and not along that perpendicular to it. This can then be taken to imply a transition of the BEC phase to a stripe phase.

B. $\theta > \cos^{-1}\frac{1}{\sqrt{3}}$

For $\theta > \cos^{-1}\frac{1}{\sqrt{3}} = 0.955$, the short range interaction become negative , and as a consequence, the spectrum is imaginary at small momentum. So, the system has long- wavelength ($q \to 0$) instability and hence a collapse of the BEC phase.

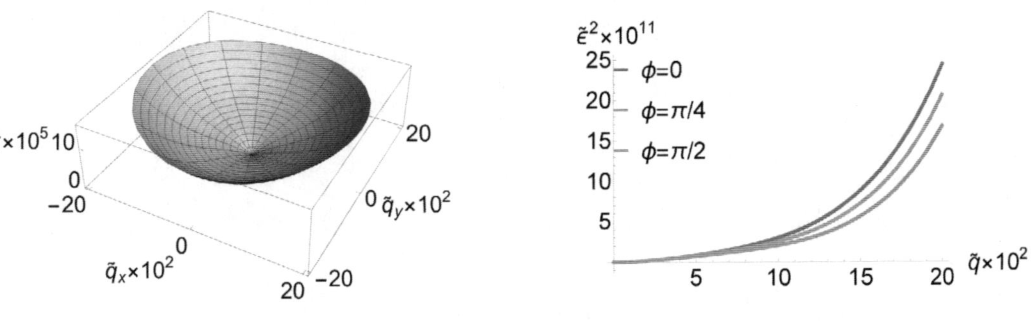

Figure 2. Figure showing the spectrum of dipolar Boson with tilt angle $\theta = 0.4$ at scaled density $\tilde{n} = 100$,. a_{dd} is taken to be $10^4 a_0$, and $r_c = 10a_0$.

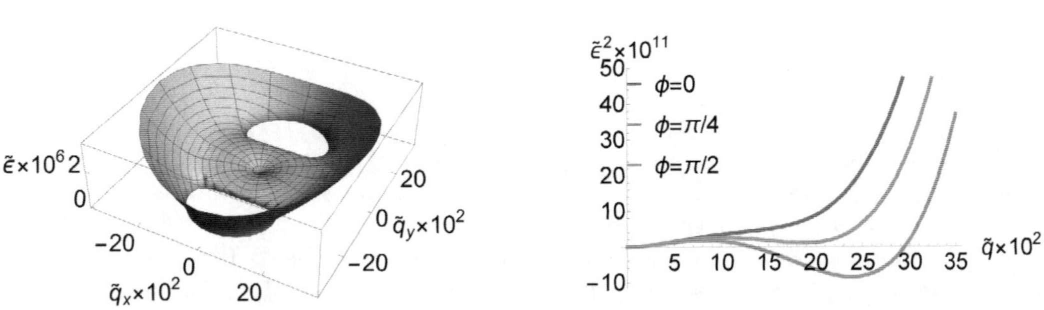

Figure 3. Figure showing the spectrum of dipolar Boson with tilt angle $\theta = 0.4$ at scaled density $\tilde{n} = 200$, a_{dd} is taken to be $10^4 a_0$, and $r_c = 10a_0$.

The set of figures in Fig. 4 are meant to show variation of the energy spectrum (ϵ^2) vs q_y with tilt angle θ, and density n. Variation with respect to density are shown for fixed $\theta = 0$, 0.4, 0.6, 0.9. The last plot in Fig. 4 shows the spectrum for $\theta = 1.0$, which is beyond the critical $\theta_c = 0.955$, and hence shows phonon instability. In each of the other figures, with increasing density, the roton minimum deepens till at sufficiently large density, the frequency becomes imaginary, shown as negative ϵ^2 in the plots.

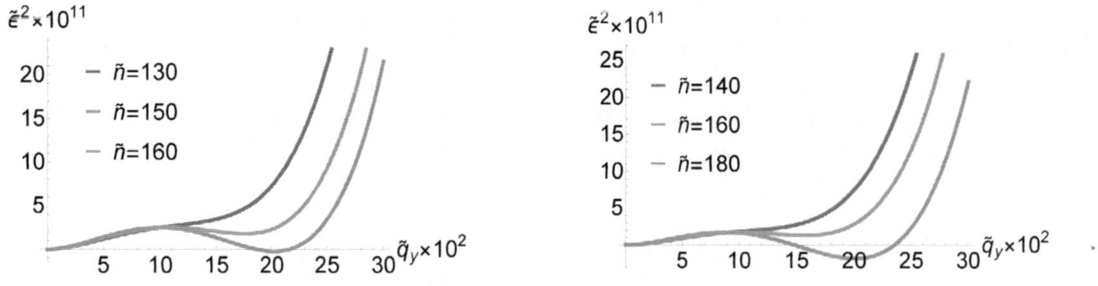

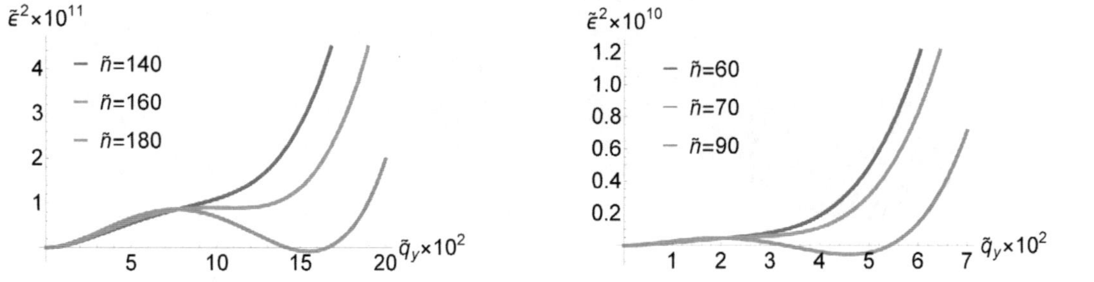

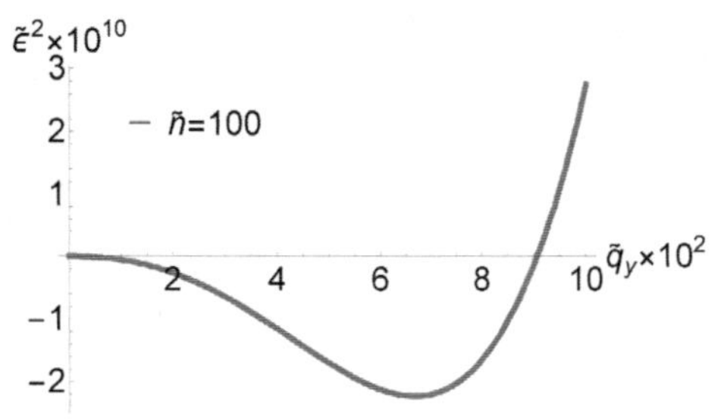

Figure 4. Figures showing the dipolar boson spectra in y direction for tilt angle $\theta = 0$ and 0.4 (first row; left and right); $\theta = 0.6$ ad 0.9 (second row; left and right). As density increases, roton minimum become imaginary. At $\theta = 1$ (bottom row), the spectrum has phonon instability.

5. Phase Diagram

We combine our results above for various tilt angles and densities to obtain a density vs tilt angle phase diagram for a homogeneous 2D dipolar boson system at zero temperature; this is shown as the left side figure in Fig 5. Since one of our goals was to see how a mean-field description compares with Quantum Monte Carlo calculation for a homogeneous 2D Bose gas subject to DDI, on the right hand plot we show phase diagram obtained from a QMC calculation [7]. It is interesting that some of the key features of the QMC phase diagram are captured in our HFB calculation. While the general phase boundary of the BEC phase, the collapse and the stripe phases are in qualitative agreement with the QMC result, we are not able to obtain the solid phase within our HFB approximation; we find the existence of density wave/stripes in the region of the QMC solid phase of high density and smallish tilt angle.

We note that our calculations are done under the assumption that the condensate density $n_0 \approx n$, which implies that the depletion from the condensate is small. Given that, we have attempted to describe the effect of a finite-range anisotropic dipole-dipole interaction in 2D, by exploring the effect of non-zero tilt angles. Our calculations and results show that in the relatively weakly interacting regime, the roton instability does not occur, and the uniform condensate is stable. As we approach the roton instability as a function of tilt angle θ and also

IOP Conf. Series: Journal of Physics: Conf. Series **1041** (2018) 012011 doi:10.1088/1742-6596/1041/1/012011

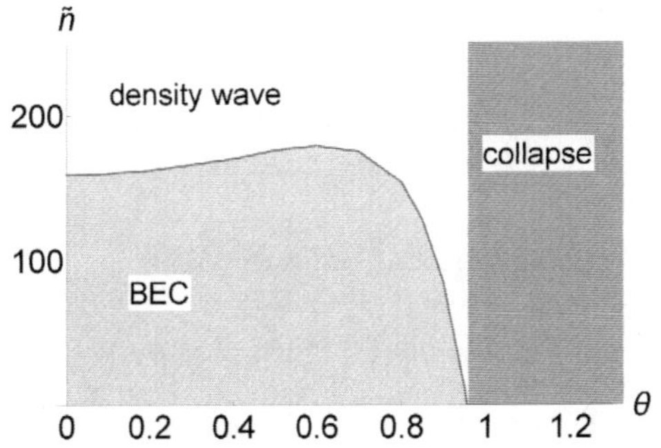

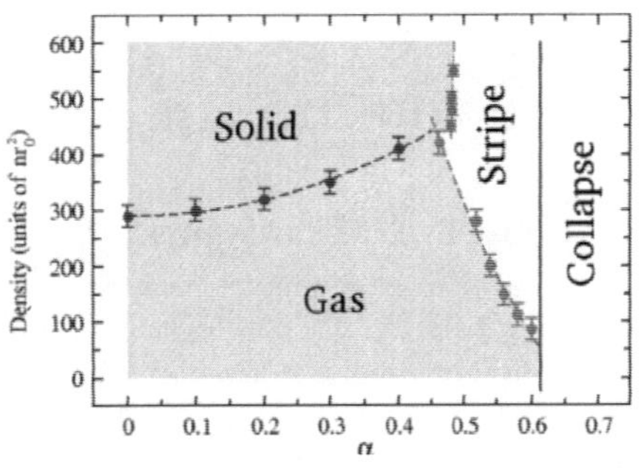

Figure 5. Scaled density vs tilt angle phase diagram. The left figure is based on our calculation; the right figure is reproduced from Ref. [7]. θ and α refer to tilt angle discussed in text. In our calculation, critical θ is determined from q-space consideration (2D Fourier-transformed $V(q)$); in Monte-Carlo calculations it is determined from r-space $V(r)$. We have chosen a_{dd} to be $10^4 a_0$, and $r_c = 10a_0$.

for large enough densities, depletion will start to increase, and at roton instability, depletion will become significant enough that the uniform condensate may not be sustainable. Since the dipolar interaction is anisotropic, it becomes attractive for some direction first, prior to becoming attractive in other directions. In our 2D case with non-zero tilt angle, this occurs first in the y-direction. Roton instability at finite momentum results in standing wave. The system exhibits density wave modulations in that direction, resulting in a "stripe phase" behavior.

Acknowledgments
We would like to thank J. Boronat, C.J. Pethick and G. Baym for useful discussions. P. Shen acknowledges support from Institute for Complex Adaptive Matter (ICAM). K. Quader acknowledges the hospitality of Aspen Center for Physics, where part of this work was done.

References
[1] Koch T, Lahaye T, Metz J, Frohlich B, Griesmaier A and Pfau T 2008 *Nat Phys* **4** 218–222 URL http://dx.doi.org/10.1038/nphys887
[2] Molony P K, Gregory P D, Ji Z, Lu B, Köppinger M P, Le Sueur C R, Blackley C L, Hutson J M and Cornish S L 2014 *Physical review letters* **113** 255301
[3] Chomaz L, van Bijnen R M, Petter D, Faraoni G, Baier S, Becher J H, Mark M J, Waechtler F, Santos L and Ferlaino F 2017 *arXiv preprint arXiv:1705.06914*
[4] Fischer U R 2006 *Phys. Rev. A* **73**(3) 031602 URL https://link.aps.org/doi/10.1103/PhysRevA.73.031602
[5] Ticknor C, Wilson R M and Bohn J L 2011 *Physical review letters* **106** 065301
[6] Macia A, Hufnagl D, Mazzanti F, Boronat J and Zillich R E 2012 *Physical review letters* **109** 235307
[7] Macia A, Boronat J and Mazzanti F 2014 *Physical Review A* **90** 061601
[8] Lange P, Krieg J and Kopietz P 2016 *Phys. Rev. A* **93**(3) 033609 URL https://link.aps.org/doi/10.1103/PhysRevA.93.033609

IOP Conf. Series: Journal of Physics: Conf. Series **1041** (2018) 012012 doi:10.1088/1742-6596/1041/1/012012

Filament Dynamics in Resistive Switching

Jong E. Han[1], Jiajun Li[1,2], Camille Aron[3,4] and Gabriel Kotliar[5]
[1] Department of Physics, University at Buffalo SUNY, Buffalo NY 14260, USA
[2] Department of Physics, Friedrich-Alexander-Universität Erlangen-Nürnberg, Erlangen 91054, Germany
[3] Laboratoire de Physique Théorique, École Normale Supérieure, CNRS, PSL Research University, Sorbonne Universités, 75005 Paris, France
[4] Instituut voor Theoretische Fysica, KU Leuven, Belgium
[5] Department of Physics, Rutgers University, New Jersey 08854, USA

E-mail: `jonghan@buffalo.edu`

Abstract. The hallmark of nonequilibrium phase transitions is the strong spatial inhomogeneity associated with bi-stability of macroscopic phases. We consider the resistive switching in transition metal oxides as a prototypical nonequilibrium phase transition associated with an equilibrium phase transition. We analyze the behavior of the hysteretic behavior of I-V relations in the resistive switching, especially the negative differential resistance regime during the reverse sweep of applied voltage. We show that the peculiar I-V behavior is closely related to the bi-stability of the metallic and insulating phases and the dynamics of the metallic filament, and investigate the filament dynamics in terms of the minimum entropy production principle in the nonequilibrium theormodynamics.

1. Introduction

Controlling materials property with electromagnetic input is one of the central themes in modern condensed matter physics. We focus here on the phase transition in transition-metal oxides controlled by static bias applied across the sample, driving the system from an insulator to metal and vice versa. This phenomenon known as the resistive switching (RS) has been studied over the last half century, and the interest on this topic has piqued recently [1, 2, 3] due to its application possibilities as a resistive logic device. Production of commercial resistive memory is imminent. The systems exhibiting the resistive transition are quite diverse – ranging from intrinsic semiconductors to transition-metal oxides/chalcoginides (TMO/TMC) – with varying mechanisms with switching fields over the range of $1 \sim 1000$ kV/cm.

The origin of the underlying RS mechanism in small-gap systems of TMO/TMC is very much debated until today. One of the main questions is the magnitude of the switching electric field for the insulator-to-metal transition (IMT) at the order of a few kV/cm, which translates to the energy scale of sub-meV of potential drop across a unit cell. From the point of the band-structure theory, this seems too small to create a fundamental transformation of electronic structure from an insulator to a metal. This raises a question of whether the appropriate approach is to consider the electric field as effectively coupled to collective degrees of freedom.

The scenarios for the RS can be divided into two categories: electronic and thermal mechanisms. The electronic scenario advocates direct action of electric field [4] through acceleration of electron kinetic energy directly impacting the electronic structure and leading to the resistive transition. Such scenarios include impact ionization, avalanche, electro-migration of metallic filaments. These mechanisms are considered fast in the regime of femto-seconds [5]. However, as pointed above, the predicted energy scales for the switching field are generally considered too large for the observed energy gap.

The other group of scenarios of thermal switching rely on the indirect role of the electric field via the hot-electron effect created by the electric field [6, 7, 8]. Electric field pumps energy into the system via coupling to electrons which then dump their excess energy to phonons. In a steady-state the heat balance is maintained between the power generated by the electric field and the energy dissipation into phonons whose energy is eventually dissipated into environment. At the nonequilibrium steady-state, the effective temperature of electrons, elevated from the bath temperature, is controlled by the electric field and a phase transition may take place due to this hot-electron temperature. This scenario has been advocated by many groups, in particular, with a result demonstrating that the hot-electron temperature at an RS in VO_2 matching the equilibrium critical temperature [6]. However, it is widely considered that the thermal process would be too slow to account for the RS phenomena and the controversies remain.

The goal of this theoretical effort is to provide a minimal dissipation model on gapped solids and reconcile the differences between the electronic and thermal RS scenarios based on a quantum mechanical microscopic model. Previous theories have largely been semi-classical resistive network model which set up a network of resistors following an empirical resistance $R(T)$ and establish local energy balance due to the Joule heating [1, 3]. This theory has been successful in reproducing the experimental I-V relations, but is fundamentally incapable to address the mechanism of the RS.

2. Filament Dynamics and Nonequilibrium Steady-State

In the past few decades, the debate in the RS phenomena has centered around the question of whether the IMT (insulaotr-to-metal transition) is electronic or thermal. Recent experiments have supported both scenarios. The experiments supporting electronic mechanism have shown that the switching mechanism is ultra-fast and is independent of phonon-pumping [5]. We have shown recently [9] that the nonequilibrium excitation by electric field is given by the Landau-Zener excitations suggesting that the switching mechanism is fundamentally electronic. However, some experiments have made arguments supporting the thermal switching ideas with the most definitive argument coming from the local temperature measurement in the vicinity of metallic filament during the switching. This work showed that the hot-electron temperature matched the equilibrium critical temperature, which has also been confirmed by our previous work. The authors have concluded that the RS in ordered solids is fundamentally an electronic process while it permits a thermal interpretation.

To put the previous theoretical findings in a broader context of thermodynamics, we examine here the scenario of nonequilibrium pattern formation based on the nonequilibrium thermodynamics argument [10] of the minimum entropy production, following Ridley's work [11]. In comparison with the numerical results, we find discrepancies from Ridley's argument to suggest that the bulk resistivity has to be significantly modified to include the spatial variation of the order parameter to reproduce the microscopic theory prediction.

Let us first review Ridley's seminal work [11] from 1963 for nonequilibrium pattern formation. The sample geometry is depicted in FIG. 1(a) where the so-called S-type material develops a metallic filament during the insulator-to-metal transition process. This is typical signature of spatial inhomogeneity in nonequilibrium phase transition. With the bulk resistivity relation *empirically* given by the resistance $R(T)$, as sketched in FIG. 1(b), between the electric field E and the current density J, the current values are given as multi-valued at given electric field such as shown. For the middle solution in the region of the negative differential resistance (NDR, $dJ/dE < 0$), the bulk solution spontaneously breaks into spatially inhomogeneous regions

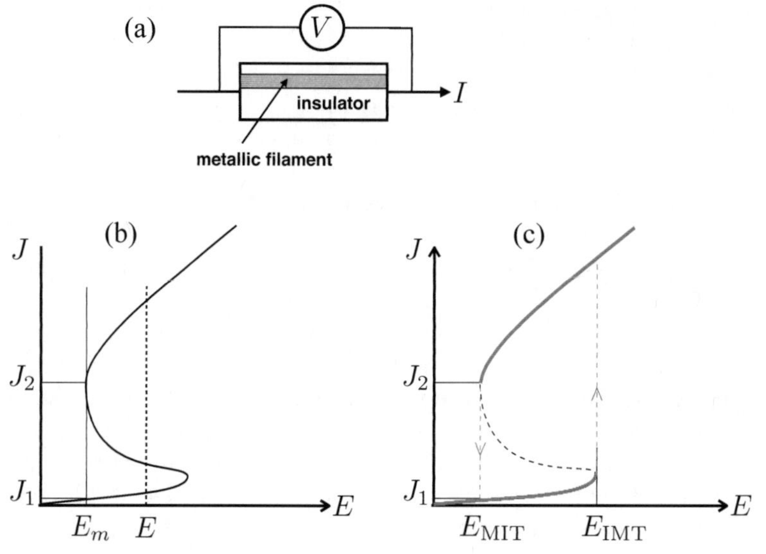

Figure 1. (a) Schematic drawing of the resistive switching devices. Given an electric current I the device supports bi stability of the insulating and metallic phases with the metallic filament carrying the majority of the current. (b) Sketch of the bulk $I-V$ relation of the S-shape RS materials. For a given I, the electric field E is determined to minimize the entropy production. For a current density J (averaged over the sample) in the range $J_1 < J < J_2$, the multiple-phase solution at $E = E_m$ minimizes the Joule heating. (c) Sketch of theoretical I-V from with a uniform order parameter. The (black) dashed part is only available in spatially inhomogeneous solution. The electric field which allows a stable uniform metallic solution is the E_{MIT} at the metal-to-insulator transition.

with metallic and insulating mixed phases [11]. The metallic phase appears in the form of a conducting filament, as shown in (a).

The criterion for the multi-phase solution relies on the *minimum entropy production principle* of the nonequilibrium thermodynamics for a steady-state. Since the S-shaped curve is single-valued as a function of the current, we imagine that we map out the $I-V$ relation of the devices by fixing the total current flowing through the device, hence the name current-controlled NDR device [12]. The nonequilibrium thermodynamics has that the entropy change of the system is written in a continuity equation of entropy S (per volume) as

$$\frac{dS}{dt} = -\nabla \cdot \mathbf{J}_s + \dot{s}, \tag{1}$$

where $\mathbf{J}_s$ is the entropy flow and $\dot{s}$ is the entropy production. $\mathbf{J}_s$ accounts for the heat flow into environment in the form of Joule heating and $\dot{s}$ is the entropy generation mainly due to the external force, and the temperature and chemical potential gradient. *Ignoring* the nonequilibrium effect of the interface between the metallic and insulating phases through the temperature and chemical potential gradient, we then express the entropy production in terms of the bulk power generation EJ. When multiple phases exist, the total EJ is the sum from each phases. At a nonequilibrium steady-state $dS/dt = 0$ and the entropy flux and the entropy production are balanced out.

The solution for the filament formation is to find the minimum entropy production given the total current I while the electric field is treated as adjustable. We also allow the fractional cross-section a of the metallic phase to vary. Since the electric field inside the metallic and insulator is identical we have the entropy production as

$$T\dot{s} = E[J_1(1-a) + J_2 a], \tag{2}$$

with the current density J_1 and J_2 inside the insulating and metallic phases, respectively. For current density in the interval $J_1 < J < J_2$ [with J_1 and J_2 defined by the NDR condition as marked as in (b)] a uniform phase is unstable, and the minimum electric field which allows multiple solutions is E_m as shown in FIG. 1(b). At $E = E_m$, J_1 and J_2 are read off from the curve $R(T)$. By imposing the constraint

$$I/A = J_1(1-a) + J_2 a, \tag{3}$$

with the device cross-section A, one can determine the filament cross-section a.

Let us examine this scenario for the filament formation based on the numerical microscopic calculations performed on finite lattices of dissipative tight-binding model with anti-ferromagnetic (AFM) ordering. The details of the model is described in Ref. [9]. The dissipation is introduced by the fermion bath model [13, 14] and the mean-field approximation is used to treat the anti-ferromagnetic correlation. Once the Hamiltonian is set, we have used the Keldysh Green's function method to solve for the self-consistent condition for AFM order parameter.

The $I-V$ characteristics as sketched in FIG. 1(c) describes the bulk calculations, with an infinite size lattice of uniform order parameter following the formalism in Ref. [15]. The red solid lines are metallic (top) and insulating (bottom) solutions. The black dashed line connecting the two phases is only accessible with mixed-phase solutions and a uniform bulk calculation is not accessible to this NDR part of the $I-V$. As the electric field is turned on from zero, the current density follows the lower curve until the threshold field $E_{\rm IMT}$ for the insulator-to-metal transition (IMT) and the system makes a strongly discontinuous transition to a metallic solution (following the red dashed line). Upon decreasing the field, the metallic state makes a transition at the field $E_{\rm MIT}$ for the metal-to-insulator transition (MIT). The calculation showed that the nature of the two transitions are quite different with the threshold field dependence,

$$E_{\rm IMT} \sim \frac{\Delta_0^2}{t} \text{ and } E_{\rm MIT} \sim \frac{\Gamma\Delta_0}{t}, \tag{4}$$

with the electron hopping parameter t, damping parameter Γ and the AFM gap Δ_0 at zero temperature and zero field. Typically the damping parameter Γ is much smaller than other energy scales and we have the relation

$$E_{\rm MIT}/\Delta_0 \ll E_{\rm IMT}/\Delta_0 \sim \Delta_0/t \ll 1. \tag{5}$$

The switching fields are much smaller than the energy scale of the energy gap, and in particular, the MIT field is orders of magnitude smaller than the IMT field.

Figure 2 shows the calculated $I-V$ relations for a square lattice oriented in the $(1,1)$ direction with respect to the electric field. The lattice of size $(60a)/\sqrt{2}\times(30a)/\sqrt{2}$ has $I-V$ relation with good qualitative agreement with experiments on VO_2 [6]. The agreement includes the resistive hysteresis, existence of the NDR in the reverse voltage sweep, and the match of the hot-electron temperature at the RS with the equilibrium critical temperature. However, calculation shows a cusp in the I-V, between the points marked as 1 and 2 in (a). This is an artifact from a finite-size calculation. As shown in (c), as one passes from 1 to 2, the insulating phase begins to appear in from the edges of the sample. The black area is for zero gap (metal) with the arrow

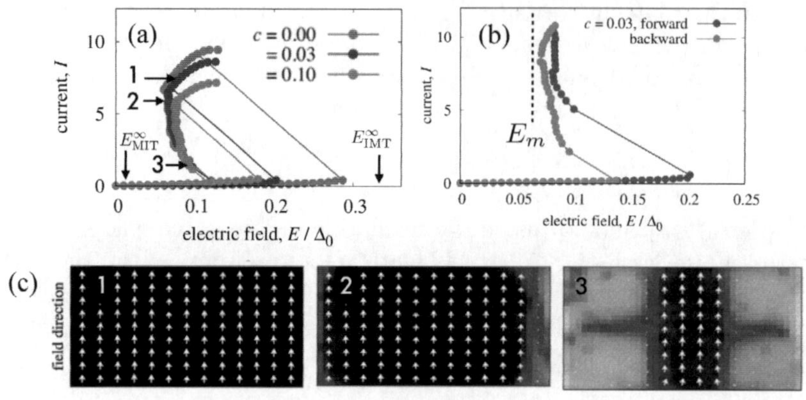

Figure 2. $I-V$ relation and the domain dynamics for resistive hysteresis in the dissipative AFM lattice. (a) $I-V$ curve for a lattice size of $(60a)/\sqrt{2} \times (30a)/\sqrt{2}$. The two-dimensional square lattice is oriented in the $(1,1)$ direction where its group velocity is aligned with the field direction. The IMT shows sensitive dependence on the impurity concentration c while the MIT is nearly independent of impurities. The IMT and MIT threshold fields calculated for an infinite and uniform lattice are marked as $E^{\infty}_{\rm IMT}$ and $E^{\infty}_{\rm MIT}$, respectively. (b) $I-V$ for a $(120a)/\sqrt{2} \times (30a)/\sqrt{2}$. The part of the curve showing the NDR behavior extends further vertically, due to more room for filament to fit in the lattice. It also shows rough steps (and slight hysteresis) in the vertical part due to the pinning of order parameters by impurities. Otherwise, the overall behavior of the hysteresis is very similar to (a). (c) Snapshots of the domain structure on the reverse sweep of total voltage in (a) at $c = 0.03$ (blue curve). The black area is for zero gap (metal) and orange area for a finite gap (insulator), and the arrows denote the current density.

showing the current density. As we move from 2 to 3, the metallic region shrink like a thinning filament. This filament dynamics is confirmed in a larger calculation with twice the width of the previous calculation with $(120a)/\sqrt{2} \times (30a)/\sqrt{2}$, as shown in (b). The nearly vertical part of the NDR curve extends further into high current without the sign of a cusp. In this run, the filament emerges with a width smaller than the lattice width. A small hysteresis loop appeared in the vertical I-V part, probably due to the pinning of domain walls by impurities. A similar behavior has been observed in V_2O_3 samples [8].

In the limit of large filament thickness, one expects that the bulk properties of the metallic and insulating domains may determine the mixed-phase properties, as argued by Ridley [11]. In the numerical results, the electric field indeed seems to approach a limit E_m, as marked in Figure 2(b). As seen in (a), the limit E_m is insensitive to the impurity concentration, and by comparing with (b) E_m does not depend strongly on the width of the system either. As the filament narrows, the I-V deviates from the vertical line and it is speculated [9] to come from inter-domain-wall scattering of electrons, leading to an increased voltage for thinner filament, i.e., smaller current.

However, the value of E_m does not agree with the value derived from Ridley's argument as discussed in Figure 1. Based on the minimum entropy production principle, E_m is expected to be the minimum electric field which stabilizes a metallic phase, that is, $E^{\infty}_{\rm MIT}$. As shown in Figure 2(a), $E^{\infty}_{\rm MIT}$ is smaller than calculated E_m roughly by an order of magnitude. As discussed above, the reason for the discrepancy is not the impurities nor the size of the system. We speculate that the problem lies in the negligence of the nonequilibrium flow of entropy due to the temperature and chemical potential, along with the energy cost due to the order parameter stiffness. It is suggestive that the domain walls grow out of the sample edges where the energy cost can be minimized. Therefore, the insulating phase may be created from the edges with lowered energy barrier.

3. Conclusion

Nonequilibrium quantum phase transition is a tremendous challenge in coming ages of condensed matter physics. We have provided some key understanding to the long-standing problem of resistive switching (RS) in ordered solids, and resolved an important issue regarding the mechanism of the RS. Quantum modeling showed that the segregation of quantum phases driven by nonequilibrium is critical to understand the nonequilibrium physics. This also indicates, however, that such calculations are going to be computationally costly, if the theory only relies on the Keldysh diagrammatics. In this work, we have tested the RS switching condition calculated from the Keldysh theory with the nonequilibrium thermodynamics argument based on the minimum entropy production principle. The discrepancy indicates that microscopic understanding of nonequilibrium entropy is essential before we begin to apply the rules of the nonequilibrium thermodynamics in quantum systems.

Acknowledgments

JL and JEH acknowledge the computational support from the CCR at University at Buffalo. JEH thanks the hospitality of the Asia-Pacific Center for Theoretical Physics.

References

[1] Janod E, Tranchant J, Corraze B, Querré M, Stoliar P, Rozenberg M, Cren T, Roditchev D, Phuoc V T, Besland M P and Cario L 2015 *Adv. Func. Mater.* **25** 6287–6305
[2] Stoliar P, Cario L, Janod E, Corraze B, Guillot-Deudon C, Salmon-Bourmand S, Guiot V, Tranchant J and Rozenberg M 2013 *Adv. Mater.* **25** 3222–3226
[3] Lee J S, Lee S and Noh T W 2015 *Appl. Phys. Rev.* **2** 031303
[4] Mayer B, Schmidt C, Grupp A, Bühler J, Oelmann J, Marvel R E, Haglund R F, Oka T, Brida D, Leitenstorfer A and Pashkin A 2015 *Phys. Rev. B* **91** 457
[5] Jager M F, Ott C, Kraus P M, Kaplan C J, Pouse W, Marvel R E, Haglund R F, Neumark D M and Leone S R 2017 *Proc. Nat. Aca. Sci.* **114** 9558–9563
[6] Zimmers A, Aigouy L, Mortier M, Sharoni A, Wang S, West K G, Ramirez J G and Schuller I K 2013 *Phys. Rev. Lett.* **110** 056601
[7] Singh S, Horrocks G, Marley P M, Shi Z, Banerjee S and Sambandamurthy G 2015 *Phys. Rev. B* **92** 155121
[8] Guénon S, Scharinger S, Wang S, Ramirez J G, Koelle D, Kleiner R and Schuller I K 2013 *EPL (Europhysics Letters)* **101** 57003
[9] Li J, Aron C, Kotliar G and Han J E 2017 *Nano Lett.* **17** 2994
[10] de Groot S R and Mazur P 1984 *Non-equilibrium Thermodynamics* (New York: Dover Publications)
[11] Ridley B K 1963 *Proc. Phys. Soc.* **82** 954
[12] Kim H T, Kim B J, Choi S, Chae B G, Lee Y W, Driscoll T, Qazilbash M M and Basov D N 2010 *J. Appl. Phys.* **107** 023702
[13] Han J E 2013 *Phys. Rev. B* **87** 085119
[14] Han J E and Li J 2013 *Phys. Rev. B* **88** 075113
[15] Li J, Aron C, Kotliar G and Han J E 2015 *Phys. Rev. Lett.* **114** 226403

IOP Conf. Series: Journal of Physics: Conf. Series **1041** (2018) 012013 doi:10.1088/1742-6596/1041/1/012013

Enhanced energy absorption through dissipation in metal clusters

M. Vincendon[1)], P. G. Reinhard[2)], E. Suraud[1,3)]

1) Laboratoire de Physique Théorique, Université P. Sabatier, 118 Rte de Narbonne, F-31062 Toulouse cedex, France

2) Institut für Theoretische Physik II, Universität Erlangen-Nürnberg, Staudstr. 7, D-91058 Erlangen, Germany

3) School of Mathematics and Physics, Queen's University Belfast, Northern Ireland, UK

E-mail: `suraud@irsamc.ups-tlse.fr`

Abstract. The paper investigates the dissipation in the electronic dynamics of metallic clusters using a theoretical description in terms of time-dependent mean-field theory with an extension to include electron-electron collisions beyond mean field in terms of a relaxation-time approximation (RTA). The topic is demonstrated on two examples: first, the damping of the all-dominant surface plasmon mode in metallic clusters becoming manifest in the spectral width of the mode and in its lifetime (when viewed in the time domain), and second, the influence of dissipation on the energy absorption of the cluster from an external laser field. The example of laser excitation reveals a dramatic enhancement of energy absorption because dissipation suppresses significantly induced photon emission which is else-wise a process limiting energy intake. The enhancement effect becomes the more important the longer the duration of the laser pulse.

1. Introduction

The optical response of metallic clusters and nano-particles is dominated by the Mie surface plasmon [1, 2, 3]. It constitutes the doorway to laser excitations and thus plays the leading role in nearly all dynamical processes [4]. The average position of the Mie surface plasmon which is an ideal collective dipole mode can be rather well described already with simple approaches as a dipole sum-rule estimate [3], dielectric theory [2], or a fluid dynamical model [5]. This suffices for many estimates in the area of clusters or nano-particles. However, the width of the Mie surface plasmon or its detailed spectral distribution is often also important and it carries a great deal of interesting information. The width characterizes the lifetime of a plasmon and can thus be associated with dissipation. In the time domain, dissipation shows up, e.g., in the damping of the dipole signal with a certain relaxation time. It is the aim of this contribution to discuss, from a theoretical perspective, effects of dissipation in the electronic dynamics of metal clusters as observed in optical absorption spectra, in time evolution of multipole moments, and in energy absorption from an intense laser field. A proper description of the plasmon's width or even detailed spectral distribution requires a deeper microscopic modeling than mere dielectric or fluid dynamical theory. We will take as basis of the microscopic description time-dependent density functional theory (TDDFT) [6, 7]. It has been applied successfully in a great variety of fields [8] amongst them also intensely in cluster physics [9]. Although , as we will see,

IOP Conf. Series: Journal of Physics: Conf. Series **1041** (2018) 012013 doi:10.1088/1742-6596/1041/1/012013

TDDFT contains already a large part of dissipation, it misses the contributions from two-body collisions. The latter are brought in by the relaxation-time approximation (RTA) which recently had been made applicable to finite electronic systems [10]. We use this RTA to investigate systematically and compare different electronic relaxation processes in small Na clusters taking Na_9^+ and Na_{40} as examples. The theoretical and numerical background is summarized briefly in section 2. Subsection 2.1, in particular, contains an overview of the various dissipation mechanisms in a cluster or molecule where we work out the distinction between one-body and two-body dissipation. The practical examples are presented and discussed in section 3. It starts with a manifestation in the width of the plasmon resonance seen in the frequency domain, continues with an analysis of relaxation times visualized in the time domain, and elaborates extensively on dissipative effects in the energy absorption during laser-driven cluster dynamics.

2. Theoretical framework

2.1. The concept of dissipation

A complete, fully microscopic description of the dynamics of a many-body system shows no dissipation in the sense that it does not loose any information. One may revert all velocities at a certain instant of time and the system evolves back to the initial state. Dissipation comes into play with reduced description and/or reduced observation. A prominent example is the fluid dynamical description of a system. It employs only local density and current as dynamical variables and thus ignores all intrinsic degrees of freedom which, in turn, leads to various sources of friction [11]. A dynamical mean-field theory, as the time-dependent density-functional theory which we will use later on, is a reduced description of a many-body system in terms of single particle (s.p.) wavefunctions $\varphi_\alpha(\mathbf{r},t)$. As such it is a fully microscopic description and time reversal invariant thus carrying no dissipation (e.g., a pure mean-field state has single particle entropy zero [10, 12]). At that level, dissipation sneaks in with dynamical correlations beyond mean field. However, when looking at a small set of observables, e.g. a couple of multipole moments, one sees dissipation already from dynamical mean-field theory. This is a case of dissipation by reduced observation. In the following, we will discuss mainly the time evolution of the dipole moment and we will investigate particularly the competition between electronic one-body dissipation and two-body dissipation. The relaxation rates which we find here remain typical for the other low multipole moments.

As already mentioned, we will focus on electronic relaxation processes. Before carrying on, we will briefly summarize all relaxation mechanisms which contribute to cluster dynamics and we do that for the actual example of relaxation rates in dipole dynamics, or widths of the dipole resonance respectively:

1) Dipole oscillations are dominated by the Mie surface plasmon resonance which is the doorway for all dipole excitations processes [2].

2) The fastest relaxation process is Landau dissipation, in spectra also coined Landau fragmentation. This is the leading one-body dissipation which emerges from the distribution of the incoming dipole strength over the energetically close $1ph$ states. This one-body process starts immediately and leads to relaxation rates for Landau damping of typically 5-20 fs. Most of the $1ph$ states reside in the regime of bound states and the excitation energy remains in the system.

3) A part of the $1ph$ states has the particle state above the ionization potential (IP). Coupling to these states leads to direct electron emission associated with similar rates of few 10 fs. Energy is carried away by the emitted electron thus cooling the system besides its typical dissipation effect of damping dipole oscillations.

4) Two-body dissipation stems from electron-electron collisions beyond mean field dynamics. It transfers excitation energy from the $1ph$ channel into higher order intrinsic excitation thus leading to thermalization of the system. The relaxation rate for this two-body process depends

sensitively on excitation energy, or temperature respectively [13]. It is usually somewhat slower than one-body dissipation with rates within 20-100 fs, but may become as fast as one-body dissipation for very energetic processes. All intrinsic excitation remains initially in the bound regime and leads to thermal electron evaporation only much later, at μs time scale.
5) Coupling to ionic motion (phonons) represents a further dissipation channel. This, however, is a slow process at ionic time scale (order of ps). 6) Ionic disorder, mostly of thermal origin, delivers two sources of dissipation. First, disorder has an impact on the electronic s.p. spectra and thus on one-body dissipation. This is a coherent process at the time scale of electronic one-body dissipation, although the thermal fluctuations of ionic structure go very slowly. Second, a laser beam which hits a thermal ensemble of clusters produces a statistical ensemble of slightly different responses which when added up incoherently yield a substantial broadening of the optical response. This broadening depends, of course, strongly on the temperature of the ensemble.

In the following, we will concentrate on electronic relaxation processes 1–4.

2.2. Mean-field and beyond

Structure and dynamics of the valence electrons of the Na cluster are described in fully quantum-mechanical fashion. Starting point is a mean-field description in terms of density functional theory. Dynamics is treated at the level of the Time-Dependent Local-Density Approximation (TDLDA) treated in the real time domain [6, 7]. The actual electronic exchange-correlation energy functional is taken from Perdew and Wang [14]. A correct description of the ionization potentials (IP) is crucial to simulate electron emission properly. Thus we augment (TD)LDA by a Self-Interaction Correction (SIC) approximated by average-density SIC (ADSIC) [15] which yields correct IP for a wide variety of atoms and molecules [16]. The coupling to the ionic cores is mediated by soft local pseudopotentials [17]. As the present study focuses on the interplay of various electronic dissipation mechanisms, we keep the ionic configuration frozen in the dynamical calculations. In practice, the leading contribution from ionic effects to the spectral widths comes from thermal fluctuations of the ions [18, 19]. These overlay the electronic contributions to the width and so would complicate the analysis.

To provide the basis for later extension, we summarize briefly the formalities of mean-field dynamics. The s.p. wavefunctions are determined by the time-dependent Kohn-Sham equations $(\mathrm{i}\partial_t - \hat{h}[\varrho])\varphi_\alpha = 0$ where the Kohn-Sham hamiltonian $\hat{h}$ is an effective one-body operator depending on the actual local density $\varrho(\mathbf{r},t)$ [20, 6, 7]. Dissipation leads to mixed quantum states and these are most efficiently described by the one-body density operator $\hat{\rho} = \sum_\alpha |\varphi_\alpha\rangle w_\alpha \langle\varphi_\alpha|$ with occupation weights $0 \leq w_\alpha \leq 1$. The time-dependent Kohn-Sham equation in terms of density operators reads $\mathrm{i}\partial_t\hat{\rho} - [\hat{h},\hat{\rho}] = 0$. It changes the s.p. wavefunctions in $\hat{\rho}$, but leaves the occupation weights w_α invariant.

Dynamical correlations beyond mean field can deliver substantial contributions to dissipation. We describe them by the relaxation-time approximation (RTA) [21, 13] which had been adapted recently to describe finite electronic systems [10, 12]. The idea is to simulate the electron-electron collisions by a relaxation to the local-instantaneous equilibrium state which is the mean field state with minimal free energy at given local density, local current, and average energy. A fraction of this equilibrium state is mixed with the actual mean-field state according to the relaxation rate which is derived from the in-medium electron-electron cross section at given density and excitation [22, 23]. In most compact form the RTA equations reads

$$\mathrm{i}\partial_t\hat{\rho} - [\hat{h},\hat{\rho}] = \frac{1}{\tau_{\mathrm{relax}}}\left(\hat{\rho} - \hat{\rho}_{\mathrm{eq}}[\varrho,\mathbf{j},E]\right) \quad , \quad \frac{\hbar}{\tau_{\mathrm{relax}}} = 0.40\frac{\sigma_{ee}}{r_s^2}\frac{E^*_{\mathrm{intr}}}{N} \quad , \tag{1}$$

where $\hat{\rho}_{\mathrm{eq}}[\varrho,\mathbf{j},E]$ is the instantaneous equilibrium for given local density $\varrho(\mathbf{r},t)$, local current distribution $\mathbf{j}(\mathbf{r},t)$, and energy E. The ingredients for the relaxation time τ_{relax} are: E^*_{intr} the intrinsic (thermal) energy of the system, N the actual number of electrons, σ_{ee} the in-medium electron-electron cross section, and $r_s = (3/(4\pi\overline{\varrho}))^{2/3}$ the Wigner-Seitz radius of the electron cloud [10, 12] computed from the system's average density $\overline{\varrho}$. The expensive part in the RTA equation is the evaluation of the local equilibrium $\hat{\rho}_{\rm eq}[\varrho, \mathbf{j}, E]$. We take advantage of the fact that dissipation is slow as compared to the fastest time scales covered in TDLDA and propagate a couple (50–100) of pure TDLDA steps before carrying out the next evaluation of the r.h.s. in eq. (1). For more details on RTA and its practical handling see [10, 12].

The stationary and time-dependent Kohn-Sham equations for mean field and single-electron wave functions are solved with standard techniques [24, 9]. The Kohn-Sham potential is handled in the Cylindrically Averaged Pseudo-potential Scheme (CAPS) [25, 26] which is an efficient and reliable approximation for metal clusters close to axial symmetry as the present test case Na_9^+ and Na_{40} are. Wavefunctions and fields are thus represented on a 2D cylindrical grid in coordinate space [27]. In our actual calculations, we employ a numerical box extending to 104 a_0 in radial direction and to 208 a_0 along the z-axis with a grid spacing of 0.8 a_0. We use a large set of s.p. states to supply sufficient space for thermalization. It reaches to about 2.7 eV above the chemical potential covering up to 168 s.p. states with maximal azimuthal angular momentum $l_z = 6\hbar$. To solve the (time-dependent) Kohn-Sham equations for the single particle (s.p.) wavefunctions, we use accelerated gradient iterations for the stationary solution [28] and time-splitting for time propagation [29]. The Coulomb field is computed with successive over-relaxation [27]. We use absorbing boundary conditions [24, 30], which gently absorb all outgoing electron flow reaching the boundaries of the grid. This serves to avoid artifact from electrons re-scattered into the active zone around the cluster. Moreover, proper absorption at the boundaries allows to compute all observables related to electron emission [31].

2.3. Excitation mechanisms and observable

The simplest excitation mechanism is an instantaneous dipole boost where each s.p. wavefunction is augmented by the same velocity field as $\phi_\alpha \rightarrow \exp(-\mathrm{i}p_0\,\hat{d}_z)\phi_\alpha$ [24, 9]. The boost momentum p_0 regulates its strength. We are characterizing the boost strength in terms of the initial excitation energy $E_0^* = Np_0^2/(2m)$ caused by the boost. The instantaneous dipole boost models to a good approximation the time-dependent Coulomb field at the cluster site for collisions with very fast ion passing by the cluster [32, 31] or extremely short laser pulses.

Most experiments use excitation by laser pulses. In our simulations, the external laser field is described as a classical electro-magnetic wave in the limit of long wavelengths. This augments the Kohn-Sham Hamiltonian by a time-dependent external dipole field $U_{\rm ext}(\mathbf{r}, t) = e^2\mathbf{r}\cdot\mathbf{e}_z\,E_0\,\sin(\omega_{\rm las}t)f(t)$ with the temporal envelop $f(t) = \sin^2\left(\pi\frac{t}{T_{\rm pulse}}\right)\theta(t)\theta(T_{\rm pulse} - t)$. The laser parameters are: (linear) polarization $\mathbf{e}_z$ here chosen along the symmetry axis, the peak field strength E_0 related to laser intensity as $I_0 \propto E_0^2$, the photon frequency $\omega_{\rm las}$, and the total pulse length $T_{\rm pulse}$. The latter is related to the full width at half maximum (of intensity) as FWHM $\simeq T_{\rm pulse}/3$.

There is great variety of observables which can be computed from TDLDA and RTA. We will use here only two of them, the dipole moment and the energy absorbed from the laser field. We consider, in particular, the dipole moment along the cluster's symmetry axis (z-axis) $D(t) = D_z(t) = \int \mathrm{d}^3r\, z\, \rho_{\rm el}(\mathbf{d}, t)$. This employs the electronic density distribution $\rho_{\rm el}$ which is naturally given in density-functional theory. The dipole moment will be shown as such, i.e. as time evolution $D(t)$, and as its Fourier transform from time to frequency domain which yields the dipole strength distribution $S_D(\omega) \propto \int dt\, e^{\mathrm{i}\omega t}D(t)$ according to the recipes of spectral analysis [33]. The energy absorbed from the electro-magnetic field is computed in standard fashion as $E_{\rm abs}(t) = \int_0^t dt' \int d^3r\, \mathbf{E}(t')\cdot\mathbf{j}(\mathbf{r}, t')$, where $\mathbf{E}$ is the external electrical field and $\mathbf{j}$ the local electronic current.

IOP Conf. Series: Journal of Physics: Conf. Series **1041** (2018) 012013 doi:10.1088/1742-6596/1041/1/012013

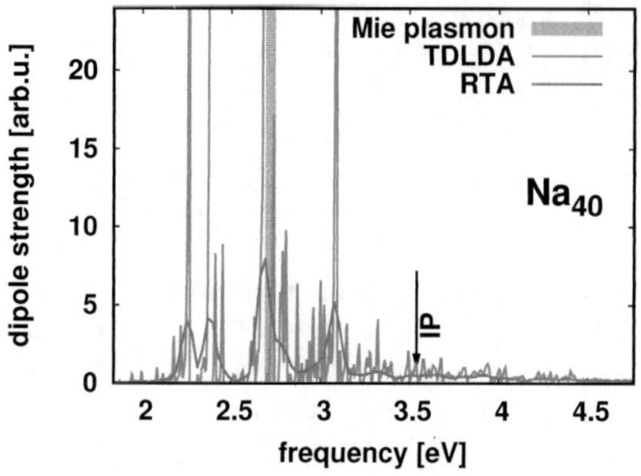

Figure 1: Distribution of dipole strength for Na_{40} computed with TDLDA and RTA in comparison. The strengths were obtained from instantaneous initial dipole boost to excitation energy $E^* = 0.3$ eV with subsequent spectral analysis [33].

3. Results

3.1. Damping mechanisms

We start by visualizing the effect of dissipation processes in spectral strength distribution. Figure 1 shows dipole strength distributions of Na_{40} at three levels of description. The blue shaded area indicates the position of the Mie surface plasmon resonance as estimated in a fluid dynamical approach [5]. This description deals only with collective flow and produces one distinct dipole resonance similar to the estimate with the dipole sum rule [3]. The next deeper level of description is mean-field theory, here in terms of TDLDA. The figure demonstrates nicely how the collective dipole strength of the Mie plasmon is distributed over the many $1ph$ state in the vicinity thus leading to a strongly fragmented distribution, called Landau fragmentation as spectral corollary of Landau damping. The electron emission threshold, the IP, is indicated in the figure. Coupling to ph states above leads to direct electron emission, the dominant energy loss channel. The step to two-body dissipation with RTA changes the spectrum once more. The global strength distribution remains unaffected. But the still spiky structure of the TDLDA spectrum is considerably smoothed thus looking closer to the spectra measured in practice. Note that we used an untypically low excitation energy to stay deep in the regime of linear response. A realistic estimate of width would have to use $E^* \approx 2.7$ eV close to the dominant plasmon resonance and then yield even smoother distributions. A proper spectral smoothing without large-scale reshuffling is the typical and most important effect of two-body correlations also in other areas of physics, see e.g. [34]. The spectrum with RTA still shows more detailed structures than experimental spectra [18]. Note that we are performing our calculations for the ionic ground state of Na_{40} kept frozen at temperature zero. The usually observed smoother spectra stem from incoherent line broadening through thermal fluctuations.

Figure 2 looks at relaxation effects in the time domain for Na_9^+ as test case. In order to avoid interference with time scales of laser excitation, we employ here excitation by an instantaneous dipole boost. All excitation energy is at once deposited on the clusters and after that pure cluster dynamics without external agents follows. This excitation mechanism is free of any time scale and serves to set a well defined initial time. The pure mean-field propagation, marked "mean field", shows a very fast damping of the dipole signal in the initial stages. This is the Landau damping associated with distribution of the dipole strength over the $1ph$ states (see figure 1). But after the first minimum, the dipole signal returns with some amplitude. Mind that the $1ph$ spectrum covers a rather low number of degrees-of-freedom such that corresponding

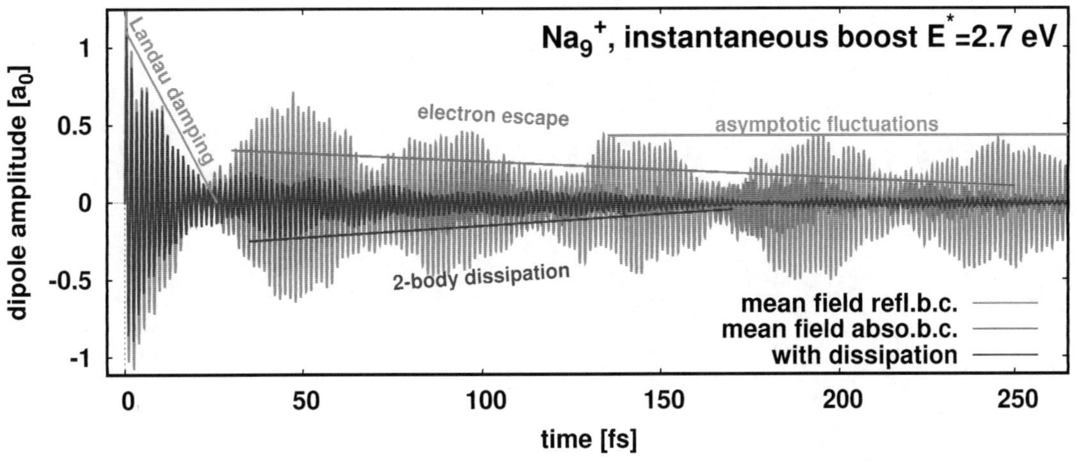

Figure 2: Time evolution of dipole momentum $D_z = \int d^3r\, z\, \rho(\mathbf{r})$ of Na_9^+ after instantaneous dipole excitation. The momentum D_z is the component along symmetry axis of the cluster.

Poincaré recurrence time comes within observation range of the simulation. The signal then shows the typical fluctuations of statistical de-coherence and re-coherence. At long times, the mean-field signal produced with physically appropriate particle emission (look here at "mean field abso.b.c.") shows a slow but steady reduction of recurrent amplitude. This signals the relaxation process from direct electron emission. These electrons are released to the continuum and never come back, thus representing true relaxation. For comparison we show also a results with reflecting boundary conditions. All electrons are forcefully kept in the numerical simulations and so add up to recurrences with undamped amplitude. Finally, we shows the result from RTA now including two-body dissipation. This substantially suppresses the recurrences and lead to a true relaxation with an asymptotically dipole signal. The figure indicates the damping time scales by straight lines associated with the rough envelopes. The hierarchy of relaxation times for the present case has Landau damping as fastest, followed by two-body dissipation, and after that comes relaxation by emission. These relations strongly depend on excitation energy. Here we have chosen $E^* = 2.7$ eV, the energy of a one-plasmon excitation. Processes at multi-plasmon energies lead to significantly faster two-body relaxation and also to faster emission. Note also that we deal here with a positively charged cluster. A neutral cluster, having lower IP, will produce more electron emission leaving less relative weight to two-body dissipation.

3.2. Dissipation versus induced photon emission

A realistic scenario for cluster dynamics is excitation by a laser pulse. The internal relaxation times within the system once excited have been studied above. The question is now which impact dissipation has on the interaction of a system with the laser pulse. A key observable for this is the energy absorbed from the external field. Figure 3 compares the time evolution of absorbed energy for TDLDA and RTA for two laser pulses of different length. The laser frequency is chosen in resonance with the strong Mie plasmon mode. We thus encounter a typical resonant scenario and we spot at first glance that dissipation makes a large difference on energy absorption. Much more energy becomes loaded from the laser field if dissipation is at work. TDLDA and RTA develop much in parallel in the initial stages. The differences develop with the first "oscillation" in $E_{\rm abs}$ which is around 50 fs for the shorter pulse (left panel) and 70 fs for the longer one (right panel). At first glance somewhat surprising, the absorbed energy suddenly turns to decrease which means that the laser field removes energy from the system

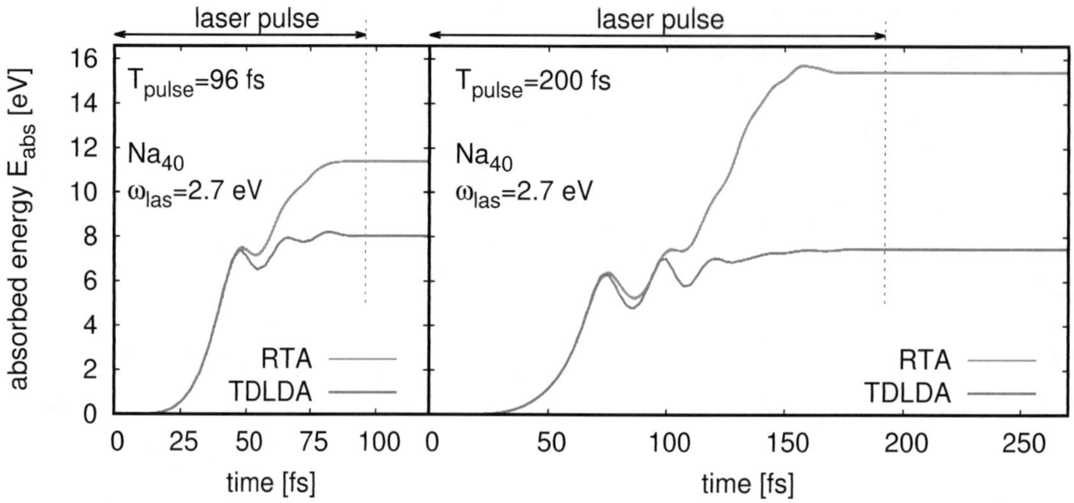

Figure 3: Time evolution of energy $E_{\rm abs}$ absorbed from the laser pulse for the Na_{40} cluster irradiated by a laser with frequency $\omega_{\rm las} = 2.7$ eV (at Mie plasmon resonance), field strength $E_0 = 0.017$ eV/a$_0$ with pulse length $T_{\rm pulse} = 96$ fs or field strength $E_0 = 0.015$ eV/a$_0$ with pulse length $T_{\rm pulse} = 192$ fs. Compared are results from TDLDA and RTA.

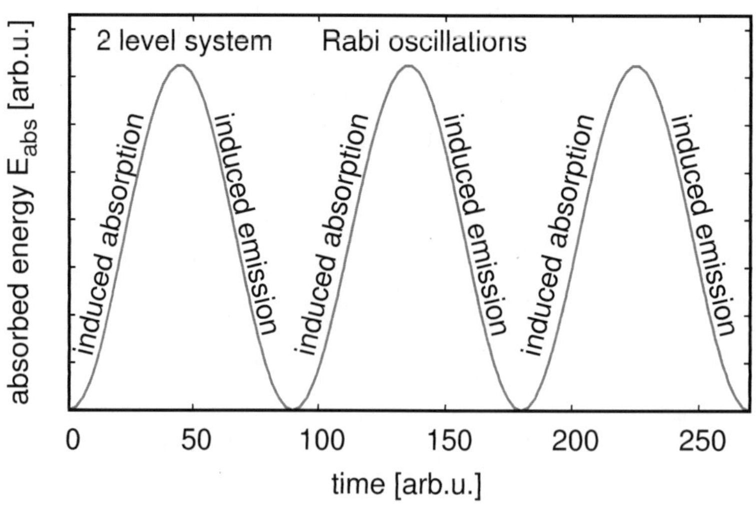

Figure 4: Schematic plot of the time evolution of energy absorption for a coherent photon field coupled to a two-level system.

rather than delivering. This is particularly pronounced for TDLDA whereas RTA suppresses this energy removal and thus allows to swallow more energy from the laser field. This repeats with each oscillation in energy absorption and eventually accumulates to a large difference in total absorbed energy, the larger the longer the pulse.

What we observe here is, in fact, a remnant of Rabi oscillations in a complex electronic system. To recall the principle, we show in figure 4 schematically the time evolution of absorbed energy $E_{\rm abs}$ for the genuine Rabi setup, a mere two-level system coupled to a monochromatic laser field. The $E_{\rm abs}$ undergoes perfect harmonic oscillations which can be explained as changing

IOP Conf. Series: Journal of Physics: Conf. Series **1041** (2018) 012013 doi:10.1088/1742-6596/1041/1/012013

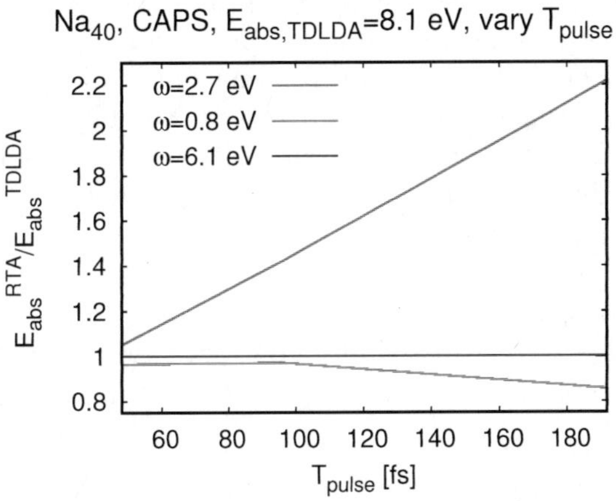

Figure 5: Ratio of final energy absorption with RTA to energy absorption with TDLDA as function of pulse length $T_{\text{pulse}}$for a laser pulse with three frequencies as indicated. The field strength, or intensity respectively, had been tuned in each case such that $E_{\text{abs}}^{(\text{TDLDA})} = 8.1$ eV for the case of TDLDA.

phases of induced photon absorption and induced emission [35], as indicated in the figure. Both processes, induced absorption and emission are always simultaneously present. But the prevalence of a process changes periodically. As soon as the photon field has sufficiently loaded the two-level system with energy, emission gains over absorption until the energy drops to a level where absorption starts to prevail. The turn to decreasing energy in laser excitation of the Na_9^+ cluster as shown in figure 3 is also caused by this effect that coherent dipole oscillations become sufficiently large to feed induced photon emission. It is, so to say, a remnant of Rabi oscillations in a complex system. It appears because the space of $1ph$ degrees-of-freedom is not large enough to destroy immediately all coherence in the dipole excitation. This changes with dissipation switched on. RTA effectively simulates the coupling to higher configurations ($2ph$, $3ph$, ...) which opens a much larger space of degrees-of-freedom, reduces coherence of the dipole oscillations, and so suppresses induced emission.

As observed in figure 3, the net effect of enhanced energy absorption becomes the larger the longer the pulse. This is demonstrated more systematically in figure 5 showing the enhancement factor for energy absorption with RTA relative to TDLDA as function of pulse length. The trend is obvious and dramatic. It demonstrates for the example of the observable E_{abs} that TDLDA becomes increasingly dubious the longer the simulations. A proper description of long-time behavior in energetic dynamical processes requires to account for dissipative effects beyond TDLDA.

4. Conclusions

We have discussed in this contribution electronic dissipative mechanisms encountered in the course of laser irradiation processes on simple metal clusters. The topic is important for numerous applications, especially in the strongly developing field of plasmonics in nano size systems. We have discussed in detail various mechanisms at work separating explicitly one-body and two body effects. While one-body effects can be analyzed within time dependent mean field theories, two body-effects (electron-electron collisions) require approaches beyond mean field. We used here a Relaxation Time Approach recently developed for finite quantum systems. Comparing RTA with mere mean-field theories, we have shown that these two-body effects play an important role for energy deposition from the laser into the system. Mean-field dynamics sooner or later hinders further energy absorption because strong coherent dipole oscillations feed enegy loss through induced photon emission. Dissipation, as modeled here by RTA, reduces dipole coherence and so paves the way to considerably enhanced energy absorption. The longer the laser pulse the larger the enhancement factor. This is an important issue which in particular points out the intrinsic limitations of time dependent mean field approaches which might strongly impact numerous applications.

Acknowledgments
This work was supported by the Institut Universitaire de France and the regional computing center Erlangen (RRZE).

[1] Mie G 1908 *Ann. Phys. (Leipzig)* **25** 377
[2] Kreibig U and Vollmer M 1993 *Optical Properties of Metal Clusters* vol 25 (Springer Series in Materials Science)
[3] Brack M 1993 *Rev. Mod. Phys.* **65** 677
[4] Fennel T, , Meiwes-Broer K H, Tiggesbäumker J, Reinhard P G, Dinh P M and Suraud E 2010 *Rev. Mod. Phys.* **82** 1793
[5] Reinhard P G, Genzken O and Brack M 1996 *Ann. Phys. (N.Y.)* **5** 576
[6] Gross E K U and Kohn W 1990 *Adv. Quant. Chem.* **21** 255
[7] Gross E K U, Dobson J F and Petersilka M 1996 *Top. Curr. Chem.* **181** 81
[8] Marques M and Gross E 2004 *Annual Review of Physical Chemistry* **55** 427–455
[9] Reinhard P G and Suraud E 2004 *Introduction to Cluster Dynamics* (New York: Wiley)
[10] Reinhard P G and Suraud E 2015 *Ann. Phys. (N.Y.)* **354** 183
[11] Balescu R 1975 *Equilibrium and Non-Equilibrium Statistical Mechanics* (New York: Wiley)
[12] Vincendon M, Suraud E and Reinhard P G 2017 *Eur. Phys. J. D* **71** 179
[13] Pines D and Nozières P 1966 *The Theory of Quantum Liquids* (New York: W A Benjamin)
[14] Perdew J P and Wang Y 1992 *Phys. Rev. B* **45** 13244
[15] Legrand C, Suraud E and Reinhard P G 2002 *J. Phys. B* **35** 1115
[16] Klüpfel P, Dinh P M, Reinhard P G and Suraud E 2013 *Phys. Rev. A* **88** 052501
[17] Kümmel S, Brack M and Reinhard P G 1999 *Eur. Phys. J. D* **9** 149
[18] Ellert C, Schmidt M, Schmitt C, Reiners T and Haberland H 1995 *Phys. Rev. Lett.* **75** 1731
[19] Montag B and Reinhard P G 1995 *Phys. Rev. B* **51** 14686
[20] Dreizler R M and Gross E K U 1990 *Density Functional Theory: An Approach to the Quantum Many-Body Problem* (Berlin: Springer-Verlag)
[21] Bhatnagar P L, Gross E P and Krock M 511 *Phys. Rev.* **94** 1954
[22] Köhn J, Redmer R, Meiwes-Broer K H and Fennel T 2008 *Phys. Rev. A* **77** 033202
[23] Köhn J, Redmer R and Fennel T 2012 *New J. Phys.* **14** 055011
[24] Calvayrac F, Reinhard P G, Suraud E and Ullrich C A 2000 *Phys. Rep.* **337** 493
[25] Montag B and Reinhard P G 1994 *Phys. Lett. A* **193** 380
[26] Montag B and Reinhard P G 1995 *Z. Phys. D* **33** 265
[27] Davies K T R and Koonin S E 1981 *Phys. Rev.* **C23** 2042–2061
[28] Blum V, Lauritsch G, Maruhn J A and Reinhard P G 1992 *J. Comp. Phys* **100** 364
[29] Feit M D, Fleck J A and Steiger A 1982 *J. Comp. Phys.* **47** 412
[30] Reinhard P G, Stevenson P D, Almehed D, Maruhn J A and Strayer M R 2006 *Phys. Rev. E* **73** 036709
[31] Wopperer P, Dinh P M, Reinhard P G and Suraud E 2015 *Phys. Rep.* **562** 1
[32] Bär M, Jakob B, Reinhard P G and Toepffer C 2006 *Phys. Rev. A* **73** 022719
[33] Calvayrac F, Reinhard P G and Suraud E 1997 *Ann. Phys. (N.Y.)* **255** 125
[34] Tselyaev V, Lyutorovich N, JSpeth, Krewald S and Reinhard P G 2016 *Phys. Rev. C* **94** 034306
[35] Milonni P W and Eberly J H 1988 *Lasers* (New York: Wiley)

IOP Conf. Series: Journal of Physics: Conf. Series **1041** (2018) 012014 doi:10.1088/1742-6596/1041/1/012014

The role of relativistic many-body theory in probing new physics beyond the standard model via the electric dipole moments of diamagnetic atoms

B. K. Sahoo

Atomic, Molecular and Optical Physics Division, Physical Research Laboratory, Navrangpura, Ahmedabad 380009, India and
State Key Laboratory of Magnetic Resonance and Atomic and Molecular Physics, Wuhan Institute of Physics and Mathematics, Chinese Academy of Sciences, Wuhan 430071, China

E-mail: bijaya@prl.res.in

B. P. Das

Department of Physics and International Education and Research Center of Science, Tokyo Institute of Technology, 2-12-1 Ookayama Meguro-ku, Tokyo 152-8550, Japan

Abstract. The observation of electric dipole moments (EDMs) in atomic systems due to parity and time-reversal violating (P,T-odd) interactions can probe new physics beyond the standard model and also provide insights into the matter-antimatter asymmetry in the Universe. The EDMs of open-shell atomic systems are sensitive to the electron EDM and the P,T-odd scalar-pseudoscalar (S-PS) semi-leptonic interaction, but the dominant contributions to the EDMs of diamagnetic atoms come from the hadronic and tensor-pseudotensor (T-PT) semi-leptonic interactions. Several diamagnetic atoms like ^{129}Xe, ^{171}Yb, ^{199}Hg, ^{223}Rn, and ^{225}Ra are candidates for the experimental search for the possible existence of EDMs, and among these ^{199}Hg has yielded the lowest limit till date. The T or CP violating coupling constants of the aforementioned interactions can be extracted by combining the EDM measurements from with atomic and nuclear calculations. In this work, we report the calculations of the EDMs of the above atoms by including both the electromagnetic and P,T-odd violating interactions simultaneously. These calculations are performed by employing relativistic many-body methods based on the random phase approximation (RPA) and the singles and doubles coupled-cluster (CCSD) method starting with the Dirac-Hartree-Fock (DHF) wave function in both cases. The differences in the results from both the methods shed light on the importance of the non-core-polarization electron correlation effects that are accounted for by the CCSD method. We also determine THE electric dipole polarizabilities of these atoms, which have computational similarities with EDMs and compare them with the available experimental and other theoretical results to assess the accuracy of our calculations.

1. Introduction

Attempts to observe an electric dipole moment (EDM) of a non-degenerate system have been steadily gaining ground for the last few decades [1, 2, 3, 4, 5, 6]. A non-zero EDM of a non-degenerate system is only possible if the system simultaneously violates parity (P) and time-reversal (T) symmetries [7, 8, 9]. In conformity with the CPT theorem [10], T violation implies charge-conjugation and parity (CP) violation. There have been direct observations of CP and T violations in the neutral K- and B-mesons [11, 12], but not in other systems. The observed CP and T violations are hadronic in nature, but it is possible that the sources producing leptonic and semi-leptonic CP violations could be responsible for the current matter-antimatter asymmetry in the Universe [13, 14]. The model dependent analysis of particle physics suggest a range of allowed CP violating interactions among elementary particles at different energy scales. The observed CP violations at lower energy scales in the K- and B-mesons have been explained by the most celebrated standard model (SM) of elementary particle physics through the Cabibbo-Kobayashi-Maskawa (CKM) matrix. However, this model predicts very small values for the EDM of the electron and the coupling coefficients for the lepton-hadron CP-violating interactions. Nonetheless, the SM is able to explain many of the observations of the fundamental properties of elementary particles, but it is unable to account for the finiteness of the neutrino masses, the matter-antimatter asymmetry of the Universe, the origin of dark matter and a few other phenomena that are of immense interest today. This leads one to believe that the SM is an intermediate manifestation of a more complete theory that needs to be uncovered. Therefore other more sophisticated extensions of the SM like the multi-Higgs, supersymmetric, left-right symmetric models have been proposed [15, 16, 17]. These models could hold the key to some of the profound unresolved questions in particle physics.

The P and T violating (P,T-odd) interactions resulting in permanent EDMs of physical systems are very sensitive to beyond the SM (BSM) physics. These interactions correspond to very high energies and are beyond the reach of the present day accelerator facilities like the large hadron collider (LHC). However, table-top EDM experiments are sensitive to CP violating effects at these high energy scales. Such experiments are currently under way in composite systems like nuclei, atoms, molecules and solids, where the interactions are strongly enhanced due to the complex dynamics of many-body effects. Many high precision measurements in atomic systems including EDMs of paramagnetic atoms and polar molecules have been reported [3, 4, 18, 19]. These systems are sensitive to a limited number of processes (e.g. leptonic, hadronic, Higgs, etc.) which arise only in selected particle physics models. In contrast, the sources of EDMs in the diamagnetic atoms can support new physics in multiple sectors of a variety of extensions of the SM. The EDMs of diamagnetic atoms arise predominantly from the electron-nucleus (e-N) tensor-pseudotensor (T-PT) interaction and the interaction of electrons with the nuclear Schiff moment (NSM) [16, 20]. The e-N T-PT interaction is the result of the CP violating electron-nucleon (e-n) interactions which arises from CP violating electron-quark (e-q) interactions at the level of elementary particles that are predicted by the leptoquark models [20]. On the other hand, the NSM can exist due to the CP violating pion-nucleon-nucleon (π-n-n) interactions, which in turn have their origins in the CP violating quark-quark (q-q) interactions or the EDMs and chromo-EDMs of quarks that are predicted by certain supersymmetric models [15, 16, 17]. In order to obtain precise limits for the coupling constants of these interactions and the EDMs of quarks, it is imperative to perform both experiments and relativistic many-body calculations reliably on atoms.

There have been remarkable advances in the past few decades in high precision measurement techniques, particularly those involving the cooling and trapping of atoms and molecules. Some of these have already been brought to bear on the measurement of EDMs in atomic and molecular systems. A breakthrough in these experiments resulting in the observation of an EDM in the foreseeable future is not beyond the realm of possibility. In fact, lowering the limits of atomic EDMs (d_a) by a few more orders would be desirable as that could discard some of the existing models of particle physics. Earlier, spin-exchange pumped masers and a ^{3}He co-magnetometer were used by Rosenberry and Chupp to yield an upper limit for the EDM of ^{129}Xe atom, $d_a(^{129}\text{Xe}) = 0.7 \pm 3.3(\text{stat}) \pm 0.1(\text{sys}) \times 10^{-27}$ e-cm [21]. Recently, new proposals to measure the EDM of this atom have been made by taking advantage of its relatively long spin relaxation time [22, 23, 24, 25, 26]. Inoue *et al.* [22] propose to utilize the nuclear spin oscillator technique [27] for carrying out the measurement of the Larmor precession lower than the above limit. On the other hand, it is anticipated that atoms like ^{223}Rn and ^{225}Ra will have large enhancements of the EDM signal owing to the large octupole deformed nuclei in these atoms [28]. Motivated by this idea, experiments to measure EDMs of these atoms are also in progress [24]. Even a measurement of the EDM of the radioactive element ^{225}Ra atom for the first time has been reported with an upper limit as $|d_a(^{225}\text{Ra})| < 5.0 \times 10^{-22}$ e-cm [2]. However, the lowest EDM limit that has been obtained to date is from ^{199}Hg and it is $|d_a(^{199}\text{Hg})| < 7.4 \times 10^{-30}$ e-cm with 95% confidence level [1].

As mentioned earlier, only the combined results of the atomic and nuclear calculations with the measured values of the EDMs of atoms are useful to constrain various CP-odd coupling constants. A number of EDM calculations have been performed on atoms that are of experimental interest, i.e. ^{129}Xe, ^{171}Yb, ^{199}Hg, ^{223}Rn, and ^{225}Ra by employing a variety of relativistic atomic many-body methods (see for more details Ref. [6]). In our recent works [29, 30, 31, 32, 33], we show that the results from the relativistic coupled-cluster (RCC) method differ by about 20-50% from the previous calculations on the aforementioned atoms that are obtained by other approximate all-order methods like the random phase approximation (RPA) [34, 35, 36], multi-configuration Dirac-Fock (MCDF) method [37, 38], a combined configuration interaction (CI) method with approximate many-body perturbation theory (MBPT) [39], and also another perturbed RCC (PRCC) method [40]. The differences between these methods have been extensively discussed by us in our earlier works [29, 30, 31, 32, 33] and the details can be found in a recent review [6]. The RPA method accounts for only the mean-field and the core-polarization effects to all-order. The coupled-cluster (CC) method, which has been referred to as the gold standard for many-body theory [41, 42, 43] has been applied to nuclear [44], atomic [45, 46], molecular [47, 48] and condensed matter systems [49, 50]. Some of the notable features of this method is that even the truncated (R)CC methods at a given level of excitation is equivalent to including correlation effects to all-orders in the residual interaction [41, 42, 47], owing to the exponential *ansatz* in expressing the wave function, it is size-extensive and size-consistent [41, 42] and amenable to high performance computational techniques [42, 47, 51, 52, 53]. It is also necessary to use relativistic theory for studying EDMs in atomic systems as EDMs are prominent only in heavy atomic systems in which relativistic effects are strong. Also, contributions from the electron EDM to the atomic EDM can only be explained by a relativistic theory [54, 55, 56, 57].

In this paper, we present our results for d_a for ^{129}Xe, ^{171}Yb, ^{199}Hg, ^{223}Rn, and ^{225}Ra atoms due to the T-PT and NSM PT-odd interactions by employing the RCC method. These values are then compared with their corresponding values from the Dirac-Hartree-Fock (DHF) and RPA methods in order to highlight the role of electron correlations for their determination and find the importance of the non-core-polarization that are neglected by RPA. We also carry out calculations of the electric dipole polarizabilities (α_d) of the above atoms using the DHF, RPA and RCC methods and compare them with the available experimental values and other theoretical calculations. The purpose of performing these calculations is the similarity of the mathematical expressions of d_a and α_d, thus the reliability of the results of the calculations of d_a can be judged to some extent by comparing the results of α_d with their measured values wherever available, or else comparing with other calculations that are reported by applying sophisticated many-body methods.

The rest of the paper is organized as follows: In the next section, we briefly mention about the theory of atomic EDMs and present the T-PT and NSM interaction Hamiltonians used for the EDM calculations. Then, we describe our many-body methods and procedures for obtaining atomic wave functions at various levels of approximations. This is followed by discussions on our results and comparison of these values with the previously performed calculations. Unless stated otherwise, we use atomic units (a.u.) throughout this paper.

IOP Conf. Series: Journal of Physics: Conf. Series **1041** (2018) 012014 doi:10.1088/1742-6596/1041/1/012014

2. Dominant sources of P,T-odd interactions

We start from the e-n P,T-odd interactions that lead to the dominant P,T-odd interactions in diamagnetic atoms. To arrive at the T-PT e-N interaction in an atom, we write the P,T-odd Lagrangian for the e-n interaction as [16]

$$\mathcal{L}_{e-n}^{PT} = C_T^{e-n}\varepsilon_{\mu\nu\alpha\beta}\bar{\psi}_e\sigma^{\mu\nu}\psi_e\bar{\psi}_n\sigma^{\alpha\beta}\psi_n + C_P^{e-n}\bar{\psi}_e\psi_e\,\bar{\psi}_n i\gamma_5\psi_n, \tag{1}$$

where $\varepsilon_{\mu\nu\alpha\beta}$ is the Levi-Civita symbol and $\sigma_{\mu\nu} = \frac{i}{2}[\gamma_\mu, \gamma_\nu]$ with the γ's being the Dirac matrices. C_T^{e-n} and C_P^{e-n} are the tensor-pseudotensor (T-PT) and scalar-pseudoscalar (S-PS) e-n interaction coupling constants respectively. Here ψ_n and ψ_e represent the field operators for the nucleon and electron respectively. This gives the T-PT e-N interaction Hamiltonian for atoms as [6]

$$H_{e-N}^{TPT} = i\sqrt{2}G_F C_T \sum_e \boldsymbol{\sigma_N}\cdot\boldsymbol{\gamma}\rho_N(r), \tag{2}$$

where G_F is the Fermi constant, C_T is the T-PT coupling constant, $\boldsymbol{\sigma_N}= \langle\sigma_N\rangle\mathbf{I}/I$ is the Pauli spinor of the nucleus with spin I, and $\rho_N(r)$ is the nuclear density.

Similarly, the Lagrangian for the other dominant P,T-odd π-n-n interaction is given by [16]

$$\mathcal{L}_{e-n}^{\pi nn} = \bar{g}_{\pi nn}^{(0)}\bar{\psi}_n\tau^i\psi_n\pi^i + \bar{g}_{\pi nn}^{(1)}\bar{\psi}_n\psi_n\pi^0 + \bar{g}_{\pi nn}^{(2)}(\bar{\psi}_n\tau^i\psi_n\pi^i - 3\bar{\psi}_n\tau^3\psi_n\pi^0) \tag{3}$$

where the couplings $\bar{g}_{\pi nn}^{(i)}$ with the superscript $i = 1, 2, 3$ represent the isospin components. This yields the corresponding e-N interaction Hamiltonian as [6]

$$H_{e-N}^{NSM} = \frac{3\mathbf{S}.\mathbf{r}}{B_4}\rho_N(r), \tag{4}$$

where $\mathbf{S} = S\frac{\mathbf{I}}{I}$ is the nuclear Schiff moment (NSM) and $B_4 = \int_0^\infty dr r^4\rho_N(r)$. The magnitude of NSM S is given by [58, 59]

$$S = g_{\pi nn} \times (a_0\bar{g}_{\pi nn}^{(0)} + a_1\bar{g}_{\pi nn}^{(1)} + a_2\bar{g}_{\pi nn}^{(2)}), \tag{5}$$

where $g_{\pi nn} \simeq 13.5$ is the CP-even π-n-n coupling constant and a_i are the polarization parameters of the nuclear charge distribution that can be computed to a reasonable accuracy using the Skyrme effective interactions in the Hartree-Fock-Bogoliubov mean-field method [58]. These couplings are related to the chromo-EDMs of up-quark ($\tilde{d}_u$) and down-quark ($\tilde{d}_d$) as $\bar{g}_{\pi nn}^{(1)} \approx 2\times 10^{-12}\times(\tilde{d}_u - \tilde{d}_d)$ [16, 60] and $\bar{g}_{\pi nn}^{(0)}/\bar{g}_{\pi nn}^{(1)} \approx 0.2\times(\tilde{d}_u+\tilde{d}_d)/(\tilde{d}_u - \tilde{d}_d)$ [16, 61], where $\tilde{d}_u$ and $\tilde{d}_d$ are scaled to 10^{-26} e-cm. Also, it yields a relation with the quantum chromodynamics (QCD) parameter ($\bar{\theta}$) by $|\bar{g}_{\pi nn}^{(1)}| = 0.018(7)\bar{\theta}$ [61]. From the nuclear calculations, one can obtain $S \simeq (1.9d_n + 0.2d_p)$ fm^2 [62]. Thus, it is necessary to obtain accurate values of C_T and S by combining atomic calculations with the experimental EDM result to extract the magnitudes of the coupling constant C_T^{e-n} and the other fundamental CP violating parameters mentioned above in a reliable manner.

3. Relativistic Atomic Many-body Methods

The EDM of the ground state $|\Psi_0\rangle$ in an atom is given by

$$d_a = \frac{\langle\Psi_0|D|\Psi_0\rangle}{\langle\Psi_0|\Psi_0\rangle}, \tag{6}$$

where D is the electric dipole moment operator. D is an odd-parity operator, so its expectation value will be zero if we consider only the electromagnetic interactions in the atoms. To obtain a non-zero EDM, the state $|\Psi_0\rangle$ must be obtained after including both the electromagnetic and weak interactions in the atomic systems. The dominant electromagnetic interactions in an atom arise due to the exchange of a single photon, and therefore we have performed our calculations in the present work in this approximation. These calculations exploit the spherical symmetry of the atoms. It would be appropriate to determine the atomic states considering first the dominant electromagnetic interactions alone and later treat the weak interaction of interest as a first-order perturbation. Consideration of both the electromagnetic and P,T-odd interactions simultaneously will mix orbitals of different parities and lead to a large configuration space for determining atomic wave functions, which require large scale computations. It is possible to reduce these computations substantially via an appropriate group symmetry for cases where the molecular orbitals are described using Cartesian coordinates. On the other hand, atomic orbitals due to the dominant electromagnetic interactions alone are described well by spherical coordinates leading to parity as a good quantum number in this case, thereby resulting in a reduction in the computational effort in performing the atomic calculations. P,T-odd interactions that have been considered in this work can be treated as first-order perturbations as their strengths are very weak.

To evaluate the atomic state functions due to the electromagnetic interactions, we consider the Dirac-Coulomb (DC) Hamiltonian which is given by

$$H^{DC} = \sum_i \Lambda_i^+ \left[c\boldsymbol{\alpha}_i \cdot \mathbf{p}_i + (\beta_i - 1)c^2 + V_N(r_i)\right] \Lambda_i^+ + \sum_{i,j\geq i} \Lambda_i^+\Lambda_j^+ \frac{1}{r_{ij}} \Lambda_i^+\Lambda_j^+, \tag{7}$$

where $\boldsymbol{\alpha}$ and β are the usual Dirac matrices, c is the speed of light, $V_N(r)$ represents for the nuclear potential and Λ^+ operators for the respective orbitals correspond to the projection operators on to positive energy levels of the Dirac orbitals. We evaluate the nuclear potential considering the Fermi-charge distribution defined by

$$\rho_N(r) = \frac{\rho_0}{1 + e^{(r-b)/a}}, \tag{8}$$

for the normalization factor ρ_0, the half-charge radius b and $a = 2.3/4(ln3)$ are related to the skin thickness. We have used $a = 2.3/4(ln3)$ and b is determined using the relation

$$b = \sqrt{\frac{5}{3}r_{rms}^2 - \frac{7}{3}a^2\pi^2} \tag{9}$$

with the root mean square (rms) charge radius of the nucleus determined using the formula

$$r_{rms} = 0.836A^{1/3} + 0.570 \tag{10}$$

in fm for the atomic mass A.

We have also included the frequency-independent Breit interaction, which is the leading relativistic correction to the two-body Coulomb interaction in our present work. This interaction is given by

$$V_B(r_{ij}) = -\frac{1}{2r_{ij}}\{\boldsymbol{\alpha}_i \cdot \boldsymbol{\alpha}_j + (\boldsymbol{\alpha}_i \cdot \mathbf{\hat{r}_{ij}})(\boldsymbol{\alpha}_\mathbf{j} \cdot \mathbf{\hat{r}_{ij}})\}. \tag{11}$$

As mentioned earlier, we only consider first-order terms in the P,T-odd weak interactions. The total atomic Hamiltonian can, therefore, be written as

$$H = H^{at} + \lambda H^{PT}, \tag{12}$$

IOP Conf. Series: Journal of Physics: Conf. Series **1041** (2018) 012014 doi:10.1088/1742-6596/1041/1/012014

where H^{at} represents the atomic Hamiltonian consisting of the DC Hamiltonian alone or with the Breit interaction added to it and λH^{PT} corresponds to either of the P,T-odd Hamiltonians given by Eqs. (2) and (4). Here λ can be S or C_T for the P,T-odd Hamiltonians. In this approach, the atomic wave function can be expanded as

$$|\Psi_0\rangle \simeq |\Psi_0^{(0)}\rangle + \lambda|\Psi_0^{(1)}\rangle \tag{13}$$

keeping terms up to first-order in λ, where $|\Psi_0^{(0)}\rangle$ and $|\Psi_0^{(1)}\rangle$ are the wave functions of H^{at} and its first-order correction due to one of the P,T-odd interaction Hamiltonians, respectively. Hence Eq. (6) can be expressed as

$$d_a = 2\lambda \frac{\langle\Psi_0^{(0)}|D|\Psi_0^{(1)}\rangle}{\langle\Psi_0^{(0)}|\Psi_0^{(0)}\rangle}. \tag{14}$$

The actual quantity that is calculated is

$$\mathcal{R} = d_a/\lambda = 2\frac{\langle\Psi_0^{(0)}|D|\Psi_0^{(1)}\rangle}{\langle\Psi_0^{(0)}|\Psi_0^{(0)}\rangle} \tag{15}$$

so that it can be combined with the experimentally measured d_a values to extract λ.

The first-order perturbed wave function $|\Psi^{(1)}\rangle$ can be calculated by two different approaches. In the sum-over-states approach, we use the expression for the first-order perturbed wave function

$$|\Psi_0^{(1)}\rangle = \sum_{I\neq 0} |\Psi_I^{(0)}\rangle \frac{\langle\Psi_I^{(0)}|H^{PT}|\Psi_0^{(0)}\rangle}{E_0^{(0)} - E_I^{(0)}} \tag{16}$$

where $|\Psi_I^{(0)}\rangle$s are the eigen states other than $|\Psi_0^{(0)}\rangle$ of H^{at} with the energies $E_I^{(0)}$ and $E_0^{(0)}$ corresponds to the ground state energy. This leads to the following expression for $\mathcal{R}$

$$\mathcal{R} = 2\sum_{I\neq 0} \frac{\langle\Psi_0^{(0)}|D|\Psi_I^{(0)}\rangle\langle\Psi_I^{(0)}|H^{PT}|\Psi_0^{(0)}\rangle}{\langle\Psi_0^{(0)}|\Psi_0^{(0)}\rangle(E_0^{(0)} - E_I^{(0)})}. \tag{17}$$

Replacing H^{PT} by D operator in the above expression will give α_d value which can be measured precisely for many atomic systems. This is the reason why electric dipole polarizabilities are calculated and compared with the available experimental data and used as a metric to ascertain the reliability of the calculated value of $\mathcal{R}$. The advantage of adopting the sum-over-states approach is that one has to calculate only the dominant low-lying state contributions and the uncertainties can be reduced by using experimental energies. However, this method cannot evaluate correlation contributions from the core and the continuum states which can be significant for heavy atomic systems. The most appropriate approach to obtain the first-order perturbed wave function involves solving the following inhomogeneous equation

$$(H^{at} - E_0^{(0)})|\Psi_0^{(1)}\rangle = (E_0^{(1)} - H^{PT})|\Psi_0^{(0)} = -H^{PT}|\Psi_0^{(0)}\rangle, \tag{18}$$

where the first-order perturbed energy $E_0^{(1)}$ vanishes since H^{PT} has odd parity. It is well known that the determination of $|\Psi_0^{(0)}\rangle$ itself accurately in heavy atomic systems is challenging because of the of two-body Coulomb and Breit interactions. So the choice of a relativistic many-body method for the evaluation of $|\Psi_0^{(0)}\rangle$ and $|\Psi_0^{(1)}\rangle$ has to be made judiciously.

IOP Conf. Series: Journal of Physics: Conf. Series **1041** (2018) 012014 doi:10.1088/1742-6596/1041/1/012014

We discuss here two variants of all-order relativistic many-body methods, RPA and RCC, to determine $\mathcal{R}$ for the aforementioned atoms that are of experimental interest, i.e. ^{129}Xe, ^{171}Yb, ^{199}Hg, ^{223}Rn, and ^{225}Ra. Earlier, relatively simple many-body methods such as the relativistic third-order many-body perturbation theory (MBPT(3) method) and the relativistic RPA were employed to determine these quantities in ^{129}Xe and ^{223}Rn [34, 35]. These methods are not sufficiently powerful for performing rigorous calculations of $\mathcal{R}$ for atoms such as ^{171}Yb, ^{199}Hg and ^{225}Ra, where pair-correlation effects contribute significantly. So the combined CI+MBPT method has been employed to evaluate $\mathcal{R}$ values in these atoms [39]. In this hybrid method, the single particle orbitals are determined using a V^{N_c-2} potential, where N_c is the total number of electrons and the electron correlation effects are accounted for by dividing the electrons into valence and core electrons.

To demonstrate the role electron correlation effects in the evaluation of $\mathcal{R}$ values, we first evaluate the DHF contribution using the expression

$$\mathcal{R} \;=\; 2\sum_{I\neq 0} \frac{\langle\Phi_0|D|\Phi_I\rangle\langle\Phi_I|H^{PT}|\Phi_0\rangle}{E_0^{DHF}-E_I^{DHF}}, \qquad (19)$$

where $|\Phi_0\rangle$ and $|\Phi_I\rangle$ are the DHF wave functions of the ground and possible excited states of an atom and $E_{0,I}^{DHF}$ are their respective energies. Since D is an one-body operator $|\Phi_I\rangle$ represents singly excited determinants with respect to $|\Phi_0\rangle$.

By perturbing the DHF orbitals to first-order and adopting a self-consistent procedure, we can express the first-order perturbed wave function due to the P,T-odd interaction as

$$|\Psi_0^{(1)}\rangle \;\approx\; \Omega_{RPA}^{(1)}|\Phi_0\rangle, \qquad (20)$$

where we define

$$\begin{aligned}\Omega_{RPA}^{(1)} &= \sum_{k}^{\infty}\sum_{p,a}\Omega_a^{p(k,1)} \\ &= \sum_{\beta=1}^{\infty}\sum_{pq,ab}\left\{\frac{[\langle pb|V(r_{ij})|aq\rangle-\langle pb|V(r_{ij})|qa\rangle]\Omega_b^{q(\beta-1,1)}}{\epsilon_a-\epsilon_p}\right\} \\ &+\sum_{\beta=1}^{\infty}\sum_{pq,ab}\left\{\frac{\Omega_b^{q(\beta-1,1)^\dagger}[\langle pq|V(r_{ij})|ab\rangle-\langle pq|V(r_{ij})|ba\rangle]}{\epsilon_a-\epsilon_p}\right\}, \end{aligned} \qquad (21)$$

with the initial guess $\Omega_a^{p(0,1)}|\Phi_0\rangle = \frac{\langle\Phi_a^p|H^{PT}|\Phi_0\rangle}{\epsilon_a-\epsilon_p}$, for the singly excited determinants $|\Phi_a^p\rangle$ with subscript a and superscript p representing single excitation configurations by replacing an occupied orbital a from $|\Phi_0\rangle$ by a virtual orbital p, ϵ stands for the single particle orbital energies and superscript k represents for orders of two-body interactions taken into account, which goes to infinity in this case. Adopting this procedure, we can evaluate $\mathcal{R}$ by

$$\mathcal{R} \;\approx\; 2\langle\Phi_0|D\Omega_{RPA}^{(1)}|\Phi_0\rangle. \qquad (22)$$

It can be shown from the above formulation that the RPA method subsumes a certain class of single excitations representing the polarization of the core by the P,T-odd interaction to all-orders. This would be the dominant correlation effect for the inert gas atoms, where the pair-correlation effects are not very significant in the all-order perturbation calculations and this is evident from the RCC calculations.

Taking $|\Phi_0\rangle$ as the reference state, $|\Psi_0^{(0)}\rangle$ is expressed in the (R)CC method as [63]

$$|\Psi_0^{(0)}\rangle = e^{T^{(0)}}|\Phi_0\rangle, \tag{23}$$

where $T^{(0)}$ is the even parity RCC excitation operator due to the residual Coulomb interaction; i.e. the difference of the two-body Coulomb and the DHF potential energies. In the particle-hole excitation formalism, we express $T^{(0)} = \sum_{I=1}^{N_c} t_I^{(0)} C_I^+$, $t_I^{(0)}$ are the amplitudes of the excitations and C_I^+ stands for a string of annihilation-creation operators corresponding to a general particle-hole excitation. The equation for the ground state of H^{at} is given by

$$H^{at}|\Psi_0^{(0)}\rangle \quad = \quad E_0^{(0)}|\Psi_0^{(0)}\rangle \tag{24}$$

with the exact ground state energy $E_0^{(0)}$. The equations for the cluster amplitudes $T^{(0)}$ are obtained by using Eq. (23) and projecting Eq. (24) on the bra state $\langle\Phi_0|C_I^- e^{-T^{(0)}}$ as

$$\langle\Phi_0|C_I^-\overline{H}_N^{at}|\Phi_0\rangle = 0, \tag{25}$$

where the de-excitation operators C_I^- are the hermitian conjugate (h.c.) of C_I^+ and $H_N^{at} = H^{at} - \langle\Phi_0|H^{at}|\Phi_0\rangle$ is the normal order Hamiltonian. We use the notation $\overline{O} = e^{-T}Oe^{T} = (Oe^{T})_l$ through out the paper for a general operator O, where the subscript l stands for the linked terms [42]. For one-body and two-body operators O, $\overline{O}$ terminates naturally [42, 47].

The inclusion of a weak P,T-odd interaction Hamiltonian or the electric dipole operator D, denoted by H_λ now onwards, will modify the ground state wave function which can be written as

$$|\Psi_0\rangle = e^{T}|\Phi_0\rangle = e^{T^{(0)}+\lambda T^{(1)}}|\Phi_0\rangle, \tag{26}$$

where the effect of the perturbation is represented by $T^{(1)} = \sum_{I=1}^{N_c} t_I^{(1)} C_I^+$ with the amplitudes $t_I^{(1)}$, which includes one-order of the weak odd-parity perturbation of interest and all-orders of the residual interaction. Here λ represents the strength of the coupling coefficient of a given P,T-odd interaction or the electric field for the evaluation of d_a and α_d, respectively. The first-order perturbed wave function can be identified as

$$|\Psi_0^{(1)}\rangle = e^{T^{(0)}}T^{(1)}|\Phi_0\rangle. \tag{27}$$

The equation for the amplitudes of $T^{(1)}$ can be obtained by solving [29, 30, 31, 32, 33]

$$\langle\Phi_0|C_I^-\left[\overline{H}_N^{at}T^{(1)} + \overline{H}_\lambda\right]|\Phi_0\rangle = 0. \tag{28}$$

For the calculations of $|\Psi_0^{(0)}\rangle$ and $|\Psi_0^{(1)}\rangle$, we consider singles (one particle-one hole) and doubles (two particle-two hole) excitations in the RCC theory (CCSD method) by restricting to $I = 1, 2$ in the amplitude equations. From the point of view of computational efficiency, we construct the intermediate diagrams of $\overline{H}_N^{at}$ by dividing it into effective one-body and two-body intermediate operators as described in detail in Refs. [6, 64, 65].

Following the above expression and expanding $|\Psi_0\rangle$, we can evaluate d_a and α_d, commonly denoted here as X, for the ground state of a closed-shell atom by [31, 42, 41, 63]

$$\begin{aligned} X &\equiv \frac{\langle\Psi_0|D|\Psi_0\rangle}{\langle\Psi_0|\Psi_0\rangle} = \langle\Psi_0|D|\Psi_0\rangle_l = \lambda\langle\Psi_0^{(1)}|D|\Psi_0^{(0)}\rangle + \langle\Psi_0^{(0)}|D|\Psi_0^{(1)}\rangle \\ &= \lambda\langle\Phi_0|e^{T^{(0)\dagger}}De^{T^{(0)}}T^{(1)} + T^{(1)\dagger}e^{T^{(0)\dagger}}De^{T^{(0)}}|\Phi_0\rangle_l \\ &= 2\lambda\langle\Phi_0|\tilde{D}^{(0)}T^{(1)}|\Phi_0\rangle_l, \end{aligned} \tag{29}$$

IOP Conf. Series: Journal of Physics: Conf. Series **1041** (2018) 012014 doi:10.1088/1742-6596/1041/1/012014

where $\tilde{D}^{(0)} = e^{T^{(0)\dagger}} D e^{T^{(0)}}$ is a non-terminating series. The terms in the series can be computed uniquely by ensuring that they do not repeat. This is achieved by first considering the linear terms and contracting D that appears in $\overline{D}^{(0)}$ with a $T^{(0)}$ or $T^{\dagger(0)}$ operator. Next, the contributions from the non-linear terms are considered and contractions are carried out with other $T^{(0)}$ and $T^{\dagger(0)}$ operators till self-consistent results are obtained.

In order to estimate dominant contribution due to the neglected triple excitations in the CCSD method, we define an excitation operator using perturbation theory in the RCC framework in the following manner

$$T_3^{(0),pert} = \frac{1}{3!} \sum_{abc,pqr} \frac{(\overline{H}_N^{at} T_2^{(0)})_{abc}^{pqr}}{\epsilon_a + \epsilon_b + \epsilon_c - \epsilon_p - \epsilon_q - \epsilon_r}, \tag{30}$$

where a, b, c and p, q, r subscripts denote for occupied and unoccupied orbitals, respectively. We denote the RCC theory with the singles, doubles and partial triples excitations approximation as the CCSD(T) method. It is pertinent to note that the triple excitation contributions coming through the $T^{(1)}$ RCC operators will be extremely small. We give the uncertainties by estimating differences between the results from the CCSD and CCSD(T) methods as the maximum contributions from the neglected higher level excitations and contributions from the high-lying orbitals that were not included in the calculations of our CCSD method.

4. Results and Discussion

In Table 1, we present α_d values and $\mathcal{R}$ due to both the T-PT and NSM interactions from our DHF, RPA, CCSD and CCSD(T) calculations for all the atoms that have been considered in this study. The polarizability values are compared with the available experimental results and other calculations employing a variety of relativistic many-body methods as mentioned in the same table. By comparing the RPA and CCSD results with the DHF values, it can be seen that the correlation trends for the individual properties of the different atoms are not similar. It can be noticed here that the RPA values for the polarizabilities of the inert gas atoms are slightly larger than the DHF values and the CCSD values increase further and are close to the experimental value for ^{199}Xe. The experimental value of this quantity for ^{223}Rn is not available but from the similarities in the trends of electron correlation effects in both the inert gas atoms, we expect that our estimated α_d of this atom is accurate and will serve as a benchmark for its measured value. In contrast, this trend is different for the polarizabilities of the other atoms where the outer two orbitals can be treated as valence orbitals and the electron correlation effects between them are very strong. The RPA values are larger than those of their CCSD counterparts. It is interesting to point out that the CCSD method implicitly contains the RPA effects. This highlights the very strong cancellations between the core-polarization effects, represented by RPA, and the non-core-polarization correlations to obtain the α_d values in these atoms. The CCSD values are also closer to the available experimental results. The correlation trends are found to be different for α_d in ^{199}Hg than they are for the other two-valence atoms ^{171}Yb and ^{225}Ra. For ^{199}Hg, the RPA value is larger than that of its DHF counterpart, but in the case of the CCSD it is the opposite, implying that non-core-polarization effects are very strong. For the other two atoms, the CCSD results lie in between the DHF and RPA values. We have also given estimates of the corrections due to the Breit interaction (ΔB) using RPA and found that they are very small. Compared to α_d , the correlation trends for d_a are found to be different. In these cases, the RPA values are always larger. The DHF and CCSD values are very close for the inert gas atoms while in other atoms both the RPA and CCSD values are substantially larger than the DHF values. Electron correlation effects contribute significantly in ^{171}Yb and ^{225}Ra compared to ^{199}Hg. From the differences between the CCSD and CCSD(T) results, it is obvious that they are important for atom. The contributions of the Breit interaction are found to be small. Since

IOP Conf. Series: Journal of Physics: Conf. Series **1041** (2018) 012014 doi:10.1088/1742-6596/1041/1/012014

the perturbations are odd-parity vector operators, the variations in the correlation trends are mainly due to the different radial behavior of these operators. The major contributions to the matrix elements of the electric dipole operator and the P,T-odd interaction Hamiltonians come from regions far away and close to the nucleus respectively. For all the quantities, we also give the "Final" values by estimating the uncertainties due to the Breit interaction, neglected triples and the finiteness of the orbital basis. These contributions are dominated by the latter two. The approximate triples are estimated using the operator defined by Eq. (30).

We also present results obtained from other methods in Table 1. Besides our RCC work, the only other method used for the EDM of the inert gas atoms is the RPA. Our RPA values agree well with those of Ref. [39]. Our CCSD results do not agree with those of other calculations because the approximations used in them are different from those made by us. The PRCC and our CCSD methods are based on the same RCC formalism, but the implementations are different in the two approaches. Unlike our approach, the normalization factor of the wave function appears explicitly in the PRCC method [40]. The CI+MBPT method uses the V^{N_c-2} DHF potential, i.e. two core electrons are excluded in generating the orbitals for the reference state, while our DHF wave function is based on the V^{N_c} DHF potential. The CI+MBPT and MCDF methods are not size-extensive unlike the RCC method. It is instructive to compare our results for polarizability of Ra with those of another RCC approach based on the numerical finite field approach [67]. Our RPA result corresponds to that of the DHF of the latter calculation, since it is carried out with the interaction of the induced electric dipole moment with the electric field in the zeroth order. The agreement between these two results is about one percent. However, the discrepancies are ten and three percents at the CCSD and CCSD(T) levels, respectively. This can be attributed to the different methods employed in the two calculations of the polarizabilities. While we use algebraic expressions arising from expectation values to perform our polarizability calculations, a two point central difference method has been used to compute the second derivatives that are needed to determine the polarizabilities in Ref. [67]. Perhaps a four point or a six point central difference computation would be necessary in order to bring the later calculation in agreement with the former. We reiterate that the results at the starting point, i.e. from the DHF method, are different in both the approaches. Therefore, both the calculations may come closer with the inclusion of full triples and quadruple excitations in this heavy atom.

To demonstrate how the correlations effects contribute through different RCC terms to our results, we give their values by classifying them into three terms as $\tilde{D}_{ob}T_1^{(1)}$, $T_1^{(0)\dagger}\tilde{D}_{ob}T_2^{(1)}$ and $T_2^{(0)\dagger}\tilde{D}_{ob}T_2^{(1)}$ along with their h.c. terms for the properties in Table 2. $\tilde{D}_{ob}$ represents the effective one-body part of the terms in the non-terminating series $\tilde{D}$ of Eq. (29) and it is determined self-consistently. By multiplying this effective one-body operator with the perturbed $T_1^{(1)}$, we get the most dominant contribution. This term includes all the RPA contributions (that also contains the lowest-order DHF results) and the other dominant non-core-polarization effects as the net single excitations. The other two terms represent contributions purely due to the non-core-polarization effects. The contributions from the later two terms are very small for d_a, but they contribute significantly to the polarizabilities. The differences between the RPA values from $\tilde{D}_{ob}T_1^{(1)}$ contributions suggest the importance of the non-core-polarization correlation effects in these studies. Therefore, it is imperative to employ a suitable many-body method that can treat all possible correlations on equal footing for reliable calculations of the EDMs and the dipole polarizabilities of the atoms that we have considered.

5. Conclusion

We have demonstrated that the RCC theory is capable of determining the EDMs and the dipole polarizabilities of diamagnetic atoms to high accuracy. The importance of accounting for non-core-polarization effects to all-orders through the RCC theory for these properties in atoms, that are neglected in RPA have been highlighted. We have also shown that the electron correlation trends for EDMs and polarizabilities differ even though operators corresponding to them are both odd-parity vector operators. This suggests that the electron correlation trends are strongly influenced by the radial behavior of the operators. Since the relativistic CCSD method is the most rigorous of all the methods we have employed to evaluate EDMs, it would be appropriate to combine the results based on it with those of the measured EDMs of diamagnetic atoms to obtain CP violating coupling constants in order to probe new physics beyond the standard model.

Acknowledgments

B. K. S. acknowledges financial support from Chinese Academy of Science (CAS) through the PIFI fellowship under the project number 2017VMB0023 and partly by the TDP project of Physical Research Laboratory (PRL). Computations were carried out using Vikram-100 HPC cluster of PRL, Ahmedabad, India.

References

[1] Graner B, Chen Y, Lindahl E G and Heckel B R 2016 *Phys. Rev. Lett.* **116** 161601
[2] Parker R H et al. 2015 *Phys. Rev. Lett.* **114** 233002
[3] Hudson J J *et al* 2011 *Nature* **473** 493
[4] Baron J *et al* 2014 *Science* **343** 269
[5] Chupp T, Fierlinger P, Ramsey-Musolf M and Singh J *Preprint* Physics.atom-ph/1710.02504
[6] Yamanaka N, Sahoo B K, Yoshinaga N, Sato T, Asahi K and Das B P 2017 *Eur. Phys. J.* A **53** 54
[7] Landau L 1957 *Nucl. Phys.* **3** 127
[8] Ramsey N F 1982 *Annu. Rev. Nucl. Part. Sci.* **32** 211
[9] Fortson N, Sandars P and Barr S 2003 *Phys. Today* **56** 33
[10] Lüders G 2000 *Ann. Phys. (N.Y.)* **281** 1004
[11] Christenson J H, Cronin J W, Fitch V L and Turlay R 1964 *Phys. Rev. Lett.* **13** 138
[12] Aaij R *et al* 2013 *Phys. Rev. Lett.* **110** 221601
[13] Canetti L, Drewes M and Shaposhnikov M 2012 *New J. Phys.* **14** 095012
[14] Dine M and Kusenko A 2003 *Rev. Mod. Phys.* **76** 1
[15] Barr S M 1993 *Int. J. Mod. Phys.* A **8** 209
[16] Pospelov M and Ritz A 2005 *Ann. Phys. (N.Y.)* **318** 119
[17] Fukuyama T 2012 *Int. J. Mod. Phys.* A **27** 1230015
[18] Regan B C *et al* 2002 *Phys. Rev. Lett.* 88 071805
[19] Murthy S A, Krause D, Li Jr Z L and Hunter L R 1989 *Phys. Rev. Lett.* **63** 965
[20] Barr S M 1992 *Phys. Rev.* D **45** 4148
[21] Rosenberry M A and Chupp T E 2001 *Phys. Rev. Lett.* **86** 22
[22] Inoue T *et al* 2013 *Hyperfine Interact.* **220** 59
[23] Furukawa T *et al* 2011 *J. Phys. Conf. Ser.* **312** 102005
[24] Rand E T *et al* 2011 *J. Phys. Conf. Ser.* **312** 102013
[25] Kuchler F *et al* 2016 *Hyperfine Interact.* **237** 1
[26] Heil W, Gemmel C, Karpuk S, Sobolev Y, Tullney K, Allmendinger F, Schmidt U, Burghoff M, Kilian W, Grüneberg S K, Schnabel A, Seifert F and Trahms L, 2013 *Annalen der Physik* **525** 539
[27] Yoshimi A *et al* 2012 *Phys. Lett.* A **376** 1924
[28] Auerbach N, Flambaum V V and Spevak V 1996 *Phys. Rev. Lett.* **76** 4316
[29] Singh Y, Sahoo B K and Das B P 2014 *Phys. Rev.* A **89** 030502(R); 2014 *Phys. Rev.* A **90** 039903(E)
[30] Sahoo B K, Singh Y and Das B P 2014 *Phys. Rev.* A **90** 050501(R)
[31] Singh Y and Sahoo B K, 2015 *Phys. Rev.* A **91** 030501(R)
[32] Singh Y and Sahoo B K 2015 *Phys. Rev.* A **92** 022502
[33] Sahoo B K and Singh Y 2017 *Phys. Rev.* A **95** 062514
[34] Martensson-Pendrill A M 1985 *Phys. Rev. Lett.* **54** 1153
[35] Dzuba V A, Flambaum V V, Ginges J S M and Kozlov M G 2002 *Phys. Rev.* A **66** 012111
[36] Latha K V P and Amjith P R 2013 *Phys. Rev.* A **87** 022509
[37] Singh A D, Das B P, Perger W F, Samal M K and Geetha K P 2001 *J. Phys.* B **34** 3089
[38] Radziute L, Gaigalas G, Jönsson P and Bieron J 2016 *Phys. Rev.* A **93** 062508

IOP Conf. Series: Journal of Physics: Conf. Series **1041** (2018) 012014 doi:10.1088/1742-6596/1041/1/012014

[39] Dzuba V A, Flambaum V V and Porsev S G 2009 *Phys. Rev.* A **80** 032120
[40] Latha K V P, Angom D, Das B P and Mukherjee D 2009 *Phys. Rev. Lett.* **103** 083001; 2015 *Phys. Rev. Lett.* **115** 059902
[41] Bishop R F 1998 *Microscopic Quantum Many-Body Theories and their Applications*, Lecture Series in Physics (Berlin: Springer)
[42] Shavitt I and Bartlett R J 2009 *Many-body methods in Chemistry and Physics* (UK: Cambidge University Press)
[43] Lindgren I and Morrison J 1986 *Atomic Many-Body Theory*, Second Edition (Berlin: Springer)
[44] Hagen G, Papenbrock T, Hjorth-Jensen M and Dean D J 2014 *Rep. Prog. Phys.* **77** 096302
[45] DeYonker N J and Peterson K A 2013 *J. Chem. Phys.* **138** 164312
[46] Nataraj H S, Sahoo B K, Das B P and Mukherjee D 2008 *Phys. Rev. Lett.* **101** 033002
[47] Crawford T D and Schaefer H F 2000 *Rev. Comp. Chem.* **14**, 33
[48] Prasannaa V S, Vutha A C, Abe M and Das B P 2015 *Phys. Rev. Lett.* **114** 183001
[49] Bishop R F and Li P H Y 2017 *Phys. Rev.* B **95** 134414
[50] Bishop R F, Hale R G and Xian Y 1994 *Phys. Rev. Lett.* **73** 3157
[51] Kallay M and Surjan P 2001 *J. Chem. Phys.* **115** 2945
[52] Kallay M, Szalay P G and Surjan P R 2002 *J. Chem. Phys.* **117** 980
[53] Hirata S 2003 *J. Phys. Chem.* A **107** 9887
[54] Sandars P G H 1965 *Phys. Lett.* **14** 194
[55] Sandars P G H 1966 *Phys. Lett.* **22** 290
[56] Sandars P G H 1968 *J. Phys.* B **1** 511
[57] Das B P 1988 *Lecture Notes in Chemistry* **50** 411
[58] Engel J, Ramsey-Musolf M J and Kolck U van 2013 *Prog. Part. Nucl. Phys.* **71** 21
[59] Haxton W C and Henley E M 1983 *Phys. Rev. Lett.* **51** 1937
[60] Pospelov M 2002 *Phys. Lett.* B **530** 123
[61] Dekens W *et al* 2014 JHEP **1407** 069
[62] Dmitriev V F and Senkov R A 2003 *Phys. Rev. Lett.* **91** 21
[63] Cizek J 1969 *Adv. Chem. Phys.* **14** 35
[64] Singh Y, Sahoo B K and Das B P 2013 *Phys. Rev.* A **88** 062504
[65] Singh Y and Sahoo B K 2014 *Phys. Rev.* A **90** 022511
[66] Seth M, Schwerdtfeger P and Dolg M 1997 *J. Chem. Phys.* **106** 3623
[67] Borschevsky A, Pershina V, Eliav E and Kaldor U 2013 *Phys. Rev.* A **87** 022502
[68] Chattopadhyay S, Mani B K and Angom D 2012 *Phys. Rev.* A **86** 062508
[69] Hohm U and Kerl K 1990 *Mol. Phys.* 69, 819
[70] Miller T M 1996 *CRC Handbook of Chemistry and Physics*, 77th edition, edited by D. R. Lide (FL: CRC Press)
[71] Goebel D and Hohm U 1996 *J. Phys. Chem.* **100** 7710
[72] Radžiūtė L, Gaigalas G, Jönsson P and Bieroń J 2014 *Phys. Rev.* A **90** 012528

IOP Conf. Series: Journal of Physics: Conf. Series **1041** (2018) 012014 doi:10.1088/1742-6596/1041/1/012014

Table 1. Results of α_d values in ea_0^3 and $\mathcal{R}$ due to T-PT (in $10^{-20}\langle\sigma\rangle|e|cm$) and NSM (in $10^{-17}[1/|e|fm^3]|e|cm$) P,T-odd interactions for the considered diamagnetic atoms from DHF, RPA, CCSD and CCSD(T) methods. The α_d values are compared with available experimental values wherever available (else we mention NA for not available) and other theoretical calculations from different many-body methods including a variant of CCSD methods that are reported in the literature.

Method	^{129}Xe	^{171}Yb	^{199}Hg	^{223}Rn	^{225}Ra
		α_d value			
DHF	26.87	124.51	40.95	34.42	204.13
RPA	26.98	179.51	44.98	35.00	296.85
CCSD	28.13	134.82	34.51	36.60	228.68
CCSD(T)	28.37	135.59	35.03	36.95	235.68
(ΔB)	(0.04)	(0.49)	(−0.01)	(0.09)	(0.19)
Final	28.4(5)	136(5)	34.5(8)	37.0(5)	236(15)
Others					
RPA	27.7 [36]	179 [35], 176.16 [36]	44.92 [35]	35.00 [39]	291.4 [36]
CI+MBPT					229.9 [39]
CCSD(T)			34.42 [66]		242.8 [67]
PRCC	26.432 [68]		33.294 [40]	35.391 [68]	
Experiment	27.815(27) [69]	142(36) [70]	33.91(34) [71]	NA	NA
		$\mathcal{R}$ due to T-PT			
DHF	0.45	−0.71	−2.39	4.48	−3.46
RPA	0.56	−3.39	−5.89	5.40	−16.66
CCSD	0.47	−2.03	−3.17	4.43	−9.81
CCSD(T)	0.474	−2.02	−3.15	4.44	−9.60
(ΔB)	(−0.01)	(0.02)	(0.03)	(−0.02)	(0.06)
Final	0.47(3)	−2.0(3)	−3.2(5)	4.4(5)	−9(1)
Others					
RPA	0.564 [36], 0.57 [39]	−3.4 [35]		5.6 [39]	−16.59 [36]
CI+MBPT			−5.1 [39]		−18 [39]
MCDF		−4.84 [38]	−4.3 [40]		
PRCC					
		$\mathcal{R}$ due to NSM			
DHF	0.29	−0.42	−1.20	2.46	−1.85
RPA	0.37	−1.91	−2.94	3.31	−8.12
CCSD	0.33	−1.49	−1.76	2.77	−6.13
CCSD(T)	0.332	−1.47	−1.75	2.78	−6.04
(ΔB)	($\sim$ 0.0)	(0.02)	(0.04)	(−0.03)	(0.06)
Final	0.33(2)	−1.5(3)	−1.8(3)	2.8(3)	−6.1(5)
Others					
RPA	0.38 [39]	−1.903 [36]		3.3 [39]	
MCDF		−2.51 [37], −2.15 [72]	−2.22 [38]		
CI+MBPT		−2.12 [39]	2.6 [39]		−8.8 [39]
PRCC			−2.46 [40]		

Table 2. Contributions to α_d values in ea_0^3 and $\mathcal{R}$ due to T-PT (in $10^{-20}\langle\sigma\rangle|e|cm$) and NSM (in $10^{-17}[1/|e|fm^3]|e|cm$) P,T-odd interactions for the considered diamagnetic atoms from different RCC and their h.c. terms of the CCSD method. Here $\tilde{D}_{ob}$ represents effective one-body operators of $\tilde{D}$, which is a non-terminating series in the CCSD method but has been evaluated adopting self-consistent approach.

RCC terms	^{129}Xe	^{171}Yb	^{199}Hg	^{223}Rn	^{225}Ra
			α_d value		
$\tilde{D}_{ob}T_1^{(1)}$+h.c.	26.672	125.217	33.034	34.748	210.845
$T_1^{(0)\dagger}\tilde{D}_{ob}T_2^{(1)}$+h.c.	0.084	0.757	0.044	0.085	1.561
$T_2^{(0)\dagger}\tilde{D}_{ob}T_2^{(1)}$+h.c.	1.370	8.842	1.429	1.772	16.276
			$\mathcal{R}$ due to T-PT		
$\tilde{D}_{ob}T_1^{(1)}$+h.c.	0.482	−2.009	−3.138	4.551	−9.637
$T_1^{(0)\dagger}\tilde{D}_{ob}T_2^{(1)}$+h.c.	−0.005	−0.021	−0.027	−0.054	−0.065
$T_2^{(0)\dagger}\tilde{D}_{ob}T_2^{(1)}$+h.c.	−0.004	0.003	0.022	−0.073	−0.108
			$\mathcal{R}$ due to NSM		
$\tilde{D}_{ob}T_1^{(1)}$+h.c.	0.325	−1.417	−1.743	2.745	−5.780
$T_1^{(0)\dagger}\tilde{D}_{ob}T_2^{(1)}$+h.c.	$\sim$ 0.0	0.014	0.008	−0.004	0.035
$T_2^{(0)\dagger}\tilde{D}_{ob}T_2^{(1)}$+h.c.	0.005	−0.088	−0.022	0.029	−0.389